A PASSION FOR SCIENCE

A PASSION FOR SCIENCE

Lewis Wolpert

Professor of Biology as Applied to Medicine,
University College London

and

Alison Richards

Producer, Science Programmes,
BBC Radio

Oxford   New York   Tokyo
OXFORD UNIVERSITY PRESS
1988

Oxford University Press, Walton Street, Oxford OX2 6DP

Oxford New York Toronto
Delhi Bombay Calcutta Madras Karachi
Petaling Jaya Singapore Hong Kong Tokyo
Nairobi Dar es Salaam Cape Town
Melbourne Auckland

and associated companies in
Berlin Ibadan

By arrangement with BBC Books, a division of BBC Enterprises Ltd

Oxford is a trade mark of Oxford University Press

Published in the United States
by Oxford University Press, New York

British Library Cataloguing in Publication Data

A Passion for science
1. Science
I. Wolpert, L.   II. Richards, Alison
509.2'2   Q162
ISBN 0-19-854213-5

Library of Congress Cataloging in Publication Data
Wolpert, L. (Lewis)
A passion for science.
1. Science—Miscellanea.   2. Scientists—Interviews.
3. Creativity in science   I. Richards, Alison.
II. Title.
Q173.W79   1988   500   88-1429
ISBN 0-19-854213-5

Typeset by Cotswold Typesetting Ltd, Gloucester
Printed in Great Britain by St Edmundsbury Press,
Bury St Edmunds, Suffolk

PREFACE

—◦◦◦◦—

These interviews were originally conducted by Lewis Wolpert and produced by Alison Richards for BBC Radio 3. We have substantially rewritten and expanded the introductions, edited the conversations, and added an opening chapter. We wish to thank the scientists for permission to publish the interviews, and Maureen Maloney for typing the manuscript.

L.W.
A.R.

CONTENTS

CHAPTER I

A PASSION FOR SCIENCE

—⇒◦◉◦⇐—

PRESENT attitudes towards science seem to indicate both ambivalence and polarization. While there is much interest and admiration for science, there is also a deep seated fear and hostility.[1] Science is perceived as materialist and dehumanizing, arrogant and dangerous. Its practitioners are a band of cold and unfeeling technicians wielding power without reponsibility. Reductionism is suspect and uncomfortable, sabotaging all the mystery and wonder of life. The threats of nuclear war and the genetic manipulation of embryos loom large.

That scientific advance may lead to good or evil ends is undeniable, but the roots of the present anti-science feeling are more tangled than this. In part they simply reflect the ideas of forbidden knowledge and retribution which are deeply embedded in western culture. More interestingly, many of the specific criticisms can be traced directly to the nineteenth century Romantic movement.[2] Coleridge was clearly an early anti-reductionist and obsessed with the difference between the living whole organism and the idea of the mechanical juxtaposition of parts: 'my mind feels as if it aches to behold and know something great, something one and indivisible'. We have, he argues 'purchased a few brilliant inventions at the loss of all communion with life and the spirit of nature'. It is the same theme which D. H. Lawrence takes up in this century. 'The Universe is dead for us, and how is it to come alive again? "Knowledge" has killed the sun, making it a ball of gas with spots; "knowledge" has killed the moon—it is a dead little earth fretted with extinct craters as with smallpox . . . The world of reason and science . . . this is the dry and sterile world the abstracted mind inhabits'. Mary Shelley's Dr Frankenstein is the epitome of the scientist unleashing forces he cannot control, and so powerful is the image that it has become part of twentieth century popular culture, shorthand and emblem of the dangers of science.

All these images come not from scientists but from poets and writers. It was Mary Shelley who created the monster, not science. And it is not just an irony that this should be so. It suggests that at least a part of the antipathy stems from the difficulties non-scientists have in understanding science. Scientists themselves must take some of the blame for this. With notable exceptions they have tended to be reluctant to explain themselves to the public at large. But the problem may go deeper. There is some evidence that the scientific mode of thought is neither natural nor comfortable. Experimental work by psychologists shows that people rely on a limited number of principles which reduce the complex task of assessing probabilities to simpler judgemental operations.[3] In a very simple example, people expect that a sequence of random events such as tossing a coin is more likely to yield H–T–H–T–H–T than H–H–H–H–H–H, which is, in fact, not the case. The studies show that thinking statistically is not only difficult for many people, but an alienating activity. As Bertrand Russell pointed out,[4] when it comes to assigning causes to events 'popular induction depends on the emotional interest of the instances, not upon their number'. Science demands one does deal with dry, statistical data, abandon basic beliefs, and perhaps accept that there is no simple linear cause. As the American literary critic Lionel Trilling observed,[5] 'This exclusion of most of us from the mode of thought which is habitually said to be the characteristic achievement of the modern age is bound to be experienced as a wound to our intellectual self-esteem.'

Nowadays it is television and radio, rather than poets and writers, who are responsible for moulding public opinion, and it is perhaps not surprising that the distorted images of science persist. Scientists are usually seen as humourless talking heads. Logico–deductive and cold, they are allowed to discourse on their findings in the well defined 'slots' allotted to science broadcasting. Although they are permitted brief appearances on the news, it is still remarkably rare to find a scientist in a soap opera or a general discussion. It is hard to escape the conclusion that those who make such programmes perceive scientists as remote from everyday life, if, indeed, they qualify as human at all. Even when science does appear in a more positive light, the stereotypes are equally familiar and equally misleading. The absent-minded and unworldly professor of the comic strip would be a liability in any self-respecting laboratory. Handsome medical heroes who can solve the problems of disease with a single brilliant experiment are, sadly, rather thin on the ground.

Part of the problem, of course, is that scientists present a false picture of themselves. As Medawar pointed out,[6] the scientific paper is a kind of fraud. The neat format with its 'Introduction' followed by 'Methods', then 'Results' and finally 'Discussion' bears no relation to the way scientists actually work. While the final results must stand up to cold and objective scrutiny, the process of achieving them rarely takes the form of the calm and logical progression suggested by the telling. Purging events of all human emotion, the formal impersonal style totally fails to indicate who actually did what and why. If the effort was a collaborative one, a whole team of widely differing individuals condenses into a faceless, characterless, Delphic oracle of science. The technicians disappear completely. So do imagination and chance, confusion and failure, and the vital conversations in the coffee room and thoughts in the bath. It is, perhaps, hardly surprising that scientists are felt to lack a certain human touch. The idea that there is such a thing as scientific method has also helped to reinforce this view. The notion that there is a formal intellectual procedure which can lead surely and inevitably to the correct conclusion has always been of much greater interest to philosophers than to scientists, but it is one that sticks in the public consciousness. There is always a sense of surprise when scientists disagree over data or the interpretation of data. Once again, their mental processes are perceived to operate in a manner quite other than those of normal people.

That this is largely untrue, I know from my own experience as a scientist. Whether or not the actual mode of thought at a technical level *is* different, doing science has a great deal in common with other kinds of intellectual and creative activity. Yet these aspects of science have been almost totally neglected. Scientists, unlike artists, are not fascinated by the process of creation. Precisely because science is concerned with the external world, and must strenuously deny the personal when it comes to judging the value of its products, there is no tradition of introspection and analysis. Whereas writers, musicians, and painters—or their biographers and critics, at any rate—think a great deal about the way they think, how the imagination is fuelled, and whence cometh their inspiration, these questions are seldom asked by, or of, scientists. It is not difficult to see why. A major difference between the arts and the sciences is, as Roszak has pointed out,[7] that in science no one goes back to the original paper. No-one, historians apart, bothers to read Einstein's, or Newton's, or Crick and Watson's original papers. Once their ideas have been accepted they become incorporated into the general body of knowledge. There is no text or work of art to contemplate, whereas with *Hamlet* or *The Last Supper,* the artefact, the creation itself, is crucial.

It would also be hard to imagine a scientist writing about his work, as Proust did, that '. . . a book is the product of a different *self* from the self we manifest in our habits, in our social life, in our vices . . . .'[8] Doing science is not about self in this sense. Nevertheless, there is no reason why one should not try to examine how scientists make their discoveries, even though those discoveries represent knowledge of a quite different kind to that embodied by art. It is inconceivable that human emotions, failings, and beliefs do not impinge upon the scientific process; that it is conducted in a vacuum, unaffected by external events. The aim of the radio programmes on which this book is based was to explore precisely this territory. It is not an exercise in the philosophy or the history of science, but an attempt to find out how individual scientists, in widely differing disciplines, go about their work: what it is like to be a mathematician as opposed to a biologist; where ideas come from; what role is played by chance and imagination. In the process we also hoped to put the popular scientific stereotypes into some kind of perspective.

Science is very much a social activity. You are always conscious of working within a community. There is the community of scientists concerned with a general field of study, such as biology. It is a large group and you relate mainly to those who work in one particular area, such as developmental biology, or genetics. You know personally, or have at least met, a large fraction of your peer group in this field. Then, as the field narrows still further, to the actual area of experiments, say chick limb morphogenesis, you are very conscious of what the other people are actually doing. You are both collaborating and competing. There is no point in competing if the others are way ahead, and it is necessary to know where they are up to in order to help your own research. That is one important reason for going to meetings. Then there is the laboratory itself. You work next to other scientists. You relate to your colleagues, the technicians who work for you, and the superiors who can control your work. In a way you try to please them all. They also often make up your out-of-lab social life.

One of the areas we wanted to explore was how different disciplines compare. Can all physicists or all mathematicians talk to each other in a common language? How different is working in Big Science to working in the kind of small group I am used to? At CERN, the European Organization for Nuclear Research in Geneva, a hundred or more people may be involved in doing a single experiment using a huge, and costly particle accelerator. We use conventional microscopes and chicken eggs. Are all fields—or all scientists—equally competitive? How important has competition been in determining the outcome of a particular piece of work or even career? There is little chance that *Hamlet* will be written twice, but once the structure of DNA has been discovered, that's it. It's too bad if you were only *almost* there.

There is an enormous amount of talk in most laboratories. On the grounds that if you cannot persuade your colleagues you won't persuade the outside world, the rule in our lab is that you must be critical of your colleagues' ideas to the point of brutality or rudeness. That's the way to test them. François Jacob and Jacques Monod, experimental biologists who won the Nobel Prize in Medicine in 1965 spoke to each other almost every day for two to three hours. Francis Crick and Sidney Brenner shared an office for twenty years. Such collaborations have helped to shape the history of science. We hoped, through the conversations, to learn more about the nature of this kind of dialogue and the role it plays in the scientific process. Often it is from talk that new ideas come. At a small meeting in Edinburgh a friend told me that he had just found that vitamin A had a particular effect on cells. This encouraged us to try a particular experiment, to rather dramatic biological effect. How often is it that a chance remark leads to an important change in direction?

Then there is the question of experimental work itself. This is very far from being either the glamorous or dangerously uncontrolled activity many people imagine. It is also a great deal more laborious and time-consuming. The ratio of results to effort is frighteningly small. It usually takes hundreds or thousands of tedious hours of work to obtain a result that can be described in a few minutes. So much of the time is spent in preparation and in waiting. Experiments in genetic engineering require following something like a very long recipe over several weeks. They require mixing solutions, pipetting solutions, centrifuging mixtures, spotting them out on gels, and in between each activity—waiting. Each step must be done in a particular time so life is determined by the clock. Then, at last, you develop a photographic plate and there, perhaps, are the results. That is if something hasn't gone wrong. If it has, then all that time has been wasted. You have to become a kind of car mechanic, trying to find out where the fault lies. Every step should have been recorded in a laboratory notebook and, hopefully, the clue will lie there. It may turn out that everything else was right, but the developer was 'off' and that ruined everything at the last step. I am always struck by the ease with which *paranormal* phenomena such as levitation and psycho-kinesis are accepted compared with the difficulty of establishing even a very simple piece of knowledge in my own field. Whereas my tiny bit of information takes many man-years, levitation, even though it invokes unmeasured forces and challenges the basis of physics, can be established and apparently accepted in man-seconds. Pseudo-science seems to me a way of getting knowledge on the cheap. The rigorous, painstaking, repetitive means by which conventional scientific knowledge has to be gained is one of the defining characteristics of science.

Not every scientist is, of course, an experimentalist. One of the topics we wished to pursue was the relationship between experimentalists and theoreticians, and whether there are parallels in the way they work. Is it possible, for example, to describe the calculations of a pure mathematician as experiments? To many scientists experimental or field work remains an important part of their life even though their main work is theoretical. Darwin spent many years working on barnacles. With others, one gets the feeling that it is the work at the bench they enjoy, and the larger issues are rather irrelevant. What is it about doing science with your own hands which is so appealing? There can be no doubt that in some ways doing experiments is like gardening—there are those with green fingers for whom everything works; for others . . . it is best if you don't work too near them. Is it possible to say anything worthwhile about what makes a good experimentalist? Does luck or serendipity play any kind of role? Often these are invoked almost as a way of minimizing scientific brilliance—it was just the wind blowing in the penicillin mould which gave Fleming his crucial cultures. But would everybody have recognized the significance of the clear areas on the dish? When Pasteur was told that he had been lucky in his research, he replied 'Fortune favours the prepared mind'. I am always struck by how the good scientists have the greatest share of luck. The vexed question of scientific method is relevant here too. Is there anything more to successful science than common sense, and the pursuit of logical internal consistency and correspondence with the external world? My own feeling is that what I do really differs very little in essence from the work of a historian; a search for explanation and connections, the process of validation or verification, the falsification of ideas. What makes the study of history different is less the approach than the subject matter. Trying to define an exclusively scientific method tends to obscure the variety of styles with which different scientists work and how different ecology may be from particle physics.

Being a scientist is partly about survival, both in the laboratory and within the larger group. Life is great fun if everyone gets on and the work is going well, but awful if there are tensions or your work is going badly, especially if that of your colleagues is going marvellously. As Confucius is claimed to have said, there is no greater pleasure than of seeing your best friend fall off the roof. The ecology of laboratory life is highly complex. Conditions are often crowded, equipment has to be shared. Close friendships develop; romance too. Sex is common among the test tubes. There is no simple way that I know to deal with people who don't get on, or get on too well, and this can be very disruptive of life in the laboratory. Even more destructive for the young scientist is friction with a supervisor. One's early training is essentially an apprenticeship: not just learning techniques but modes of thought. If the relationship is good, one can identify with particular qualities and acquire a particular style. If the relationship is not so good, a young research worker's whole future may be at stake.

Some survival strategies are common. There is a curious and distinctive brand of laboratory humour which softens and helps one cope with criticism. Murphy's law is highly respected: if anything can go wrong it will, and usually at the most unexpected stage. Certainly people in laboratories laugh a lot. But most scientists also have to find their own ways of dealing with the demands of this way of life. The thrill of success is rare. It requires a certain spiritual fortitude, quite often, simply to keep going. Many—most—of your ideas turn out to be wrong, and months of experimental work can be fruitless. There is a not unreasonable temptation to wish not to commit yourself to a programme which, if it fails, yields, after months, nothing, and leaves you no further than you were before. Pride, reputation, and confidence can all be laid on the line. My first paper in biology was a disaster. I had been working at a marine station in Millport in Scotland on cell division in the sea urchin embryo, and had published in the prestigious journal, *Nature,* that a substance known as ATP blocked division. This had exciting implications and the following summer I went to a marine station in Sweden to continue the work. To my horror I could not repeat the experiments. I soon found that I had not, in the original series, carefully enough controlled the activity of the ATP. The result was an artefact I had created myself. I had to publish a retraction and recover from quite a deep depression.

Mark Ptashne of Harvard has spoken of the psychic courage required to undertake certain lines of research. Youth certainly helps; so does the support of one's colleagues. But even this cannot always be relied upon. My best piece of work also had a depressing sequel. In the spring of 1968 I presented my new ideas on pattern formation in development at a small meeting in the glorious Villa Serbelloni on the shores of Lake Como. The theory suddenly made sense of an enormous number of different embryological results and I was very excited. The ideas were quite well received at the meeting and some of those there immediately took up some of the ideas and produced novel and testable models. That summer I was at the Wood's Hole Laboratory in the USA and gave one of the Friday evening lectures. These were very well attended—several hundred people. My ideas were met with silence and at the reception afterwards I was introduced to a leading American embryologist who at once turned his back on me. Next day no one said a word about the lecture. I asked one of my American friends what was going on. 'Well, Lewis, they're all saying, who in the hell do you think you are?' It was only Sydney Brenner who was encouraging and helped me through a very dark summer. People didn't like being told that they had been thinking about the problem in the wrong way.

Most scientists experience failure and set-backs; most get stuck; most have to face the criticisms of their colleagues. Yet, almost by definition, science is only concerned with success. Certainly its public face, whether in the form of a scientific paper or a television programme, presents only what worked, not what didn't. To explore just how much of scientific life in fact consists of failure and the antagonism of one's peers was one of our main aims when we set about talking to scientists. We also hoped to discover something of how each individual copes, and how such events have shaped their careers.

Such questions are clearly closely related to those of motivation; why someone does science in the first place; what the satisfactions and compensations are. I am suspicious of the idea that scientists do what they do from a wish to help mankind. More selfish motives must surely play a role; pleasure for example, or the desire for admiration. Certainly there is the wish to *know*. It isn't only one's own successes that are thrilling. Hearing of an advance in one's own area—but not too close—can be equally invigorating. I remember the first time I heard the paediatric surgeon Judah Folkman speak. It was such a marvellous lecture. He started by describing how he offered a holiday for two in Miami to any of his students who could grow in culture a tumour more than 2 mm in diameter. The prize was safe. The point is that for tumours to grow they need blood vessels and he was isolating the factors which seemed to attract blood vessels to the tumour. I was so excited by these quite new ideas that I was almost stopping people in the street to tell them. Perhaps it is, above all, the thrill of ideas which binds scientists together, is the passion which drives them and enables them to survive.

With passion, however, comes knowledge. What you find may be of little interest outside your chosen field, or it may have much wider implications for people's health, safety, or beliefs. Many find the concepts of evolution and neurophysiology as uncomfortable to live with as the military applications of nuclear fission. It is this which, as we pointed out at the beginning, is the source of much of the suspicion surrounding science. We wanted to know how scientists themselves regarded the power of their ideas, and the distrust it evokes. Do cosmologists have any difficulty living at peace with such ideas as the beginning of time or the immensity of space; does an evolutionist ever doubt that Darwin was right? More important, perhaps, *do* scientists pursue their intellectual ends without regard for the consequences? Do individuals draw lines in their own research? Do scientists have a greater responsibility towards society than anyone else, or would they agree with Robert Oppenheimer's distinction between pure and applied science: 'The scientist is not responsible for the laws of nature, but it is a scientist's job to find out how these laws operate. It is the scientist's job to find ways in which these laws can serve the human will. However, it is not the scientist's job to determine whether a hydrogen bomb should be used. This responsibility rests with the American people and their chosen representatives'.

The answers to these questions, as to all the others, are those of a few individuals. Whether they are satisfactory is a personal judgement. But whatever else emerges from the conversations which follow, the cold, unsmiling talking head does not.

References

1. Wolpert, L. (1987). Science and Anti-science. *J. Royal College of Physicians*, **21** 159–65.
2. Willey, B. (1934). *The seventeenth century background*. Chatto & Windus, London.
3. Kelly, H. H. (1973). *American Psychologist* **28,** 109.
4. Russell, B. (1927). *Philosophy*. Norton, New York.
5. Trilling, L. (1973). *Mind in the modern world*. Viking Press, New York.
6. Medawar, P. B. (1963). Is the scientific paper a fraud? *The Listener,* 12 September.
7. Roszak, T. (1972). *Where the wasteland ends*. Doubleday, New York.
8. Naipul, V. S. (1987). On being a writer. *New York Review of Books*, 23 April.

FIRST AND LAST THINGS

ABDUS SALAM *was born in 1926 and is Professor of Theoretical Physics at Imperial College, London, and Director of the International Centre for Theoretical Physics in Trieste.*

CHAPTER 2

SCIENCE SUBLIME

———✦———

ABDUS SALAM
Theoretical physicist

THE GREEKS wished to explain all the phenomena of Nature in terms of four elements: fire, air, earth, and water. Modern science takes the urge to simplify still further. One of the most exciting and romantic endeavours of the twentieth century is the attempt to show that what we now know to be the four fundamental forces of nature—gravity, electromagnetism, and the strong and weak nuclear forces—are aspects of a yet more basic principle. This is the province of the particle physicists, the theoreticians and experimentalists who are concerned with the behaviour of electrons, protons, and the host of other sub-atomic particles that make up matter. It is a formidable undertaking.

The two forces with which we are most familiar are gravity and electromagnetism. At the beginning of the nineteenth century electricity and magnetism were thought to be quite distinct, and it was the work of Michael Faraday and, later, James Clerk Maxwell which showed that they were basically two aspects of the same phenomenon. Maxwell's four equations provide a complete description of the production and interrelation of electricity and magnetism, and subsequently led to the development this century of what are known as the gauge theories. Gravity was, of course, first recognized by Newton, and shown by Einstein in his General Theory of Relativity to be an expression of the curved geometry of space and time. Einstein tried until the end of his life to unify electromagnetism with gravitation but, like everyone since, failed to do so.

The recognition of the existence of two more fundamental forces— the strong and the weak nuclear forces—came with the discoveries about the structure of the atom in the first decades of this century. It became clear that an atom is not an indivisible sphere like a billiard ball, but consists of a small central region—the nucleus—surrounded by a cloud of particles called electrons. The nucleus itself is composed of particles called protons and neutrons, which are bound together by the strong nuclear force. But atomic nuclei are not always stable. During certain kinds of radioactive decay, for example, the neutrons are transformed into protons, electrons, and elusive particles called neutrinos. Transformations of this kind are governed by the weak nuclear force.

In the late 1950s Abdus Salam began to address the problem of unifying the weak force and the electromagnetic force. On the face of it, the two forces are very different. One, the electromagnetic force, is a long range force which can be felt at almost any distance. The other, the weak nuclear force, is a short range force acting over unimaginably short distances. But Salam succeeded. In 1979, together with Steven Weinberg and Sheldon Glashow, he was awarded the Nobel Prize for Physics. Their work was mathematical and theoretical, but predicted that certain, as yet undiscovered, particles should exist. It was only in 1983, under the cosmic conditions created in the huge particle accelerator at CERN that these particles were detected, and the theory finally confirmed.

Salam was born in Pakistan in 1926 and went to Cambridge to read mathematics at the end of the second world war. Given his subsequent achievements, he had come to the right place. As an undergraduate he attended lectures given by the distinguished theoretical physicist Paul Dirac, who had been awarded a Nobel Prize for his work on quantum mechanics and antimatter. After taking his first degree Salam went on to do research at the Cavendish Laboratory where much of the work on the structure of matter had been done. He is now Professor of Theoretical Physics at Imperial College, London, and Director of the International Centre for Theoretical Physics in Trieste, which he founded to enable young physicists from developing countries to enjoy short, intensive periods of research and international contact. He returns frequently to Pakistan, and is a devout Moslem.

I interviewed Salam at his home in south London. We could hear children playing in the next room and the familiar sounds of domestic life in the background. I could not help but be aware of a contrast between the sheer ordinariness of our surroundings and the intellectual reach of the ideas we were to discuss. Yet in his life, as well as in his research, Salam is able to accommodate seemingly disparate elements with ease: the familiar and the arcane; a commitment to physics in the developing world and to his own work on unification; to science and to Islam. I wanted to learn something of the route which had brought him from a peasant community in Pakistan to international acclaim in particle physics.

————

I was trained by my well-wishers, and my father in particular, to think of the Civil Service as a career and it was simply an accident that I became a particle physicist. The accident of the second world war. If the war had not been going on the Indian Civil Service examination would have been held in the crucial years in which I had to make up my mind for my career and I would by now be a good Civil Servant.

'You had no idea of doing science, then?'

No. It was an accident in the following way. The war had stopped the Civil Service examination. Just after the war the examination was still stopped and having done my MA in mathematics at the University of Lahore, I was given a scholarship to read mathematics further at Cambridge.

'So you had a scientific bent from quite a young age?'

Well, there was a scientific bent all right, but the point was that I was taking mathematics not to do research, but in order to score very high marks in the Civil Service examination. It was a mark spinner if you like.

'So it wasn't that you had a passion for science?'

No. I was certainly *good* in science. In fact, I was recalling the other day that I had written my very first research paper when I was about sixteen years of age and it was published in a mathematics journal. So the research mindedness was there, but there was no motivation for it. But after two years of Cambridge, of course I was hooked on research.

'But I'm not quite sure how you got to Cambridge.'

I got to Cambridge by means of a scholarship from a Small Peasants' welfare fund which was set up by the Prime Minister of the State of Punjab at that time.

'Did you come from a peasant background?'

That's right. Although my father was a Civil Servant, he had a small parcel of land and he qualified. So I got one of those scholarships and the interesting thing is that only five scholarships were offered, and the other four people who got them could not get university admission that year. Then came the partition of the country and the scholarships disappeared. So the entire purpose of that fund and those scholarships seemed to be to get me to Cambridge.

'Did you really think that fate was playing a hand? After all, each of these events was very much a matter of chance.'

Certainly my father, who was a deeply religious man, always said that this was a result of his prayers. He wanted his son to shine in some field. Of course, in the beginning he was thinking of me as a Civil Servant, but when I decided that I was going to do research he felt that this was something very appropriate and encouraged me. But the whole sequence of events, my getting a scholarship at the right time, my getting to Cambridge at all at the right time, and then being interested in science, was all, he thought, very much a part of something deeper.

'When you got to Cambridge did you become immediately involved in theoretical physics?'

No, I started in mathematics because I had the mathematical background but, slowly and gradually, during the two years in mathematics I shifted over to theoretical physics. Dirac was lecturing at that time and I attended his lectures. Then I still had a third year free in the sense that I had the scholarship and the choice of whether to go on with higher mathematics—that's Part III of the mathematics tripos—or to do the physics tripos. One of my teachers was the astronomer Fred Hoyle, and I went to ask Fred his opinion about this. He said 'If you want to become a physicist, even a theoretical physicist, you must do the experimental course at the Cavendish. Otherwise you will never be able to look an experimental physicist in the eye'. And that was very correct advice. But it was a hard year for me doing experimental work at the Cavendish, not having done any for so long. It was the hardest year of my student days.

'What did you find so hard about it?'

The whole attitude towards experiment. It's very interesting. In the Cavendish there used to be a tradition that you were not given any fancy equipment. Just string and sealing wax. You were given every discouragement and you had to overcome this. Now the very first experiment I was asked to do was to measure the difference in wave length of the two sodium D lines, the most prominent lines in the sodium spectrum. I reckoned that if I drew a straight line on the graph paper then its intercept would give me the required quantity I wanted to measure. As you know, mathematically a straight line is defined by two points and if you take one other reading then mathematically that should be enough since you then have three points on that line, two to define the straight line and the third one to confirm. So I spent three days in setting up that equipment. After that I took three readings, and I took them to be marked. In those days the marking of experimental work in the class counted towards your final examination. Sir Denys Wilkinson, who is now Vice Chancellor of Sussex University, was one of the men who supervised our experimental work, and I took it to him. He looked at my straight line, and said 'What's your background?' I said 'Mathematics'. He said 'Ah, I thought so. You realize that instead of three readings you should have taken one thousand readings and drawn a straight line through them'. I thought 'I'll be damned if I can go back and face those three days again'. I had by that time dismantled my stuff and I didn't want to go back. So I tried very hard to avoid Sir Denys Wilkinson—he wasn't Sir at that time—during the rest of the year.

I still remember the day the results came out in 1949. I was looking at the result sheets hung in the Cavendish and Wilkinson came up behind me. He looked at me and said 'What sort of class have you got?' and I very modestly said 'Well, I've got a first class'. He turned full circle on his heel, three hundred and sixty degrees, turned completely round and said 'Shows you how wrong you can be about people'. But going back to Fred Hoyle, I think his advice was absolutely right.

'Now you got the Nobel Prize for unifying certain parts of the theory in particle physics. How did you get the idea?'

It's such an *attractive* idea. You see, the whole history of particle physics, or of physics, is one of getting down the number of concepts to as few as possible. And when you are doing this 'getting down' it seems absolutely the natural thing. In fact, it always surprises me that some of my physics friends—and some of them very eminent people, Nobel Prize winners—would not subscribe to the idea. They would find the difficulties in uniting two totally disparate looking phenomena so overwhelming that they would think you stupid to think otherwise.

'Do you think your religious views made you think that they could be unified?'

I think perhaps at the back of my mind. I wouldn't say consciously. But at the back of one's mind the unity implied by religious thought perhaps plays a role in one's thinking.

'Steve Weinberg came to the same theory quite independently. That's surprising isn't it?'

Not at all. The ideas in our subject are common and the diffusion of ideas is astonishingly large. Everybody knows almost everything of what's going on. I think it's a result of the system which we have developed of Summer Schools and symposia, and of course, the pre-print system. It's a very efficient system and we in theoretical physics are probably the best organized for that, for some curious reason. Although, mind you, at the time that Steve and I were working on the theory we were using ideas which, although they had been published, were not highly regarded. In that sense we had the field more to ourselves than would be the case today.

'Did people accept the theory at once?'

No, not at all. The theory was elaborated in 1967 and it was completely ignored. In fact, even before that, when I took another paper—one I wrote in 1964—to be published in a journal the editor said 'The thing that you are predicting has already been tested and not found. Will you add the words that this paper is purely speculative?' And I had to, in order to get the thing published. Those experiments were wrong, the ones which he was alluding to, but we only found that out later.

'So how did the theory come to be accepted?'

As I said, the theory was elaborated in 1967. There was a young physicist called t'Hooft, a Dutchman, who played a very crucial role in showing that the theory was mathematically well established. It was his first piece of work, at the age of 25 or so, and it gave the idea more respectability among the theoreticians. That was in 1971. Then in 1973 the experimenters redid those experiments which had previously shown that our ideas were not correct. They were redone properly at CERN, and that gave the first indication that the theory was on the right lines. But then the experiments were repeated in the United States and they contradicted the Geneva results. And it went on like that for a couple of years, back and forth between the experimenters.

'It's interesting that those experiments turn out to be wrong. As an outsider one thinks that the one thing you have that is reliable in physics is the experimental data. I'm surprised that the facts are so often wrong'.

Well, you see, take one experiment which is going on now and which concerns the next stage of unification. I said we have united electromagnetism with the weak nuclear force. But there's a second nuclear force, the so called strong nuclear force, and that is not yet united with the electro-weak force. We hope it will be, and most of us would like to believe that this is happening. The crucial experiment for this is the decay of the proton. The proton is supposed to be, or was supposed to be, before this theory came along, a fundamentally stable particle. This theory says no, in 10^{32} years all protons will decay. That's a very, very long time. The life of the universe as you know is 10^{10} years. So in 10^{32} years . . . my goodness . . . every proton will decay! Now to see this experimentally you need 10^{32} protons to be watched for one year before one of them will decay. At the present time the situation is that there is an Indian experiment, 7000 feet deep in the Kolar Gold Field mines, which claims to have seen three events of proton decay. There is the experiment in Japan which claims to have seen one event and there are much more statistically significant experiments in the United States which claim to have seen none. Now what do you believe? The experiments are very difficult. I don't know which way the camel will sit, but this is a crucial experiment. And so it's perfectly possible that some of the experiments are wrong, or that their interpretation is wrong, and that we have to wait for more statistically significant signals.

'Now, you are a theoretician, so you're sitting there, rather grandly, while these experimentalists are testing your theories. But for the people doing the experiments with these big machines life is rather different.

The papers they publish have fifty, even a hundred, authors. Do people mind that?'

I think many experimental physicists do not like this situation. Many of them would rather have the old days when just one man or two or three people collaborated and did an experiment and enjoyed it. But the nature of the beast is now such that you cannot help it. You have to have large collaborations because the experiments are costly, and they need vast quantities of equipment. For example, one hundred and fifty experimenters were associated with the two experiments at CERN which finally showed the validity of our theory. And the equipment is incredible. The detecting devices are three storeys high.

'Is your field a very competitive one?'

Oh yes. The practitioners number of the order of about five thousand in theory and about the same number in experiment. And then there is a premium on youth as you know very well.

'Why? Do you just think you're better when you're young?'

No, you are less encumbered. You don't live with your past. You don't live with your failures. You are much more willing to try more ideas in a different way. Older people are also, of course, more encumbered with various types of administrative duties in order to keep the subject running and that sort of thing. But I think most of all it's being unencumbered with the past ideas which you have tried to use and failed. Because then you think 'Oh the idea is dead' when it's only the particular approach which you took to the idea which may be dead. I think the younger you are the better it is, if you can take the risk.

'Now were you young when you began to work on your unification theory?'

The idea started about 1957 when I was 31 which is fairly young, but then it took a long time in the execution.

'And did you get up every morning and work on it?'

No, no, not at all. It was intermittent. You worked on that particular set of ideas, then you gave up and took up something else and then you came back to it and so on, publishing little bits as you went along.

'But were you ever wrong? Have you ever been wrong in a major sort of way?'

It's probably just egotism but I can't think of anything where one has been proved completely wrong. There have been many stupid ideas which have led nowhere, of course, but that's the fate of all of us. The majority of our ideas, 99 per cent, lead nowhere. You're lucky if *one* of your ideas is correct in the end.

'And you have no misgivings about that?'

Not at all. But I think in our subject when you look at the successful ideas, you feel there is an inevitability about them. The only word I can use is 'sleepwalking'. *The Sleepwalkers* is the title of Arthur Koestler's book about Copernicus, Kepler, and Galileo. You just are led more or less from one step to the next.

'Sleepwalking seems a very passive way of doing physics''

It's sleepwalking in the following good sense. The unification ideas needed what we call the gauge theories. The gauge theories were actually first discovered in 1879 by Maxwell—as suggested by the equations for electromagnetism he had written down—then clarified by the German mathematician Hermann Weyl in 1929. They were put into the form in which we use them today by Yang and Mills and my pupil Shaw in 1954. It was the same old set of ideas which had started with Maxwell in 1879 but put in a slightly larger context. Then we, that is Weinberg, Glashow, and myself, said 'These gauge ideas are the ideas that we need.' That was our contribution. Newton, you remember, was asked why he was so great and he said 'I was not so great. I was standing on the shoulders of giants'. So my own feeling is that in each generation there is a set of ideas which is more or less common, but people forget and ascribe the entire success of those ideas to the one man who makes the best use of them. In that sense, maybe, physics has always been sleepwalking. When I said that in 1879 Maxwell had a great idea—well, he had inherited a set of ideas from Faraday. He wrote down Faraday's equations and found they were inconsistent, so he supplied one extra term. So in that sense it's inevitable, it's sleepwalking. Take Einstein's ideas which we consider the most revolutionary, the ideas of the curvature of space and time which explain gravitation. They go back to the German mathematician, Gauss, who first, in fact, made the tests to determine the curvature of space. What he didn't do was to put time into it. So there's an inevitability about these ideas. Although it was an act of genius for Maxwell to have found that extra term, and for Einstein to have added time to three-dimensional space, if you trace the history of the ideas they go back by gradations to earlier and earlier generations.

'Do you think if there hadn't been these geniuses these steps would have been made anyhow?'

Yes I do.

'Now, you come from a religious background, was there ever any conflict in doing physics?'

No, why should there be? Because fortunately, and I think I have said this so often in my writings, Islam is one of the three religions, which emphasize the phenomena of nature, and their study. One eighth of the Koran is exhortation to the believers to study nature and to find the signs of God in the phenomena of nature. So Islam has no conflict with science.

'What is the pleasure that you got from physics?'

Well let's put it this way. When you go to sleep and you are exhausted after a day's administrative chores or whatever, what is the thought that gives you maximum relief? I don't know what it is for you, but I get my pleasure from thinking about the problems of physics. It gives me the biggest relaxation.

'You mean thinking about physics is not work for you?'

It's a pleasure. I should qualify this by saying that when you are working something through it can be very, very hard and you eat your heart out. You think this idea should work and it doesn't. Then it can be devastatingly worrisome. But otherwise, most of the time you're thinking about it, it's a pleasure.

'What is that pleasure? Is it the pleasure of reflecting on what you've done that day, or contemplating the beauty of physics?'

Well after you have found something it's marvellous.

'So it's success which gives you pleasure?'

It's not just success. When you are relaxing, as I described, you are not reflecting on the successes of the past. In fact, any paper which you write does not give pleasure for more than a few days. I think for a week at the most. For a week you may be euphoric. 'Oh yes, that was a marvellous result.' But then it becomes just part of you. Presumably it becomes a part of your pleasure-giving cells, wherever they are, and impels you to further work.

'Do you still feel a sense of awe at the extraordinary nature of particle physics?'

Yes. It is always incredible that what people work out actually does happen.

'But are you more impressed by what people work out, or by what the nature of Nature is?'

Both. As a phenomenon, take Brain Science, for example. It is marvellous. So in that sense physics is not unique. But when I think in terms of the sublime theories that come in physics, I think that is unique.

'Do you like music? I mean, do you get the same sense listening to music?'

I would not say that I find the same sublimity. I find the same sublimity in reading or listening to the Koran, because there I find, for example, after you've been listening to it for half an hour, you suddenly get caught in an elevating fashion.

'But you do see physics as sublime?'

Yes, yes, no question about it. I mean take Einstein's theory—you still, after so many years, you still think 'what a sublime, what a marvellous idea it is!'

MARTIN REES *was born in 1942 and is
Plumian Professor of Astronomy and
Experimental Philosophy at
Cambridge University.*

CHAPTER 3

CONTEMPLATING THE COSMOS

MARTIN REES
Cosmologist

AT THE other end of the scale from particle physics is cosmology. Particle physics deals with the unthinkably small; cosmology with the unthinkably big. Our image of the heavens has been changing and expanding since Copernicus dethroned the earth from its central position, but even as late as 1920 it still seemed possible that our own Galaxy—the Milky Way—represented the extent of the Universe. Now we live with the knowledge that we are only one of a local group of galaxies, which is itself only one of innumerable such groups, receding outwards into space.

This scale of things is almost impossible to grasp. For most of us, the size of our astronomical backyard is problem enough. A light year—the distance travelled by light in one year—is six million million miles, and our galaxy is 100 000 light years across. Yet cosmologists, who spend their days in trying to describe the past, present, and future of the entire Universe, grapple with distances not just of thousands, but of millions, and even billions, of light years. And just as physicists have discovered a whole menagerie of exotic particles in the past two or three decades, so the astronomers have assembled a baffling and extraordinary zoo of objects for the cosmologists to explain.

In the early 1960s, for example, came the first observations of what have come to be known as quasars—quasi-stellar objects. As their name suggests, they look like stars but, according to most interpretations, they are enormously distant and exceedingly luminous objects, whose brightness can only be accounted for by an output of energy on an unprecedented scale. On the other hand it is conceivable, though increasingly unlikely, that quasars lie much closer to home, and some extraordinary local conditions produce the phenomena we see.

Also in the 1960s came the discovery that space is filled with so-called microwave background radiation. According to the Big Bang theory, the Universe was created between ten and twenty thousand million years ago, with a giant explosion which threw out matter in all directions. At temperatures of billions of degrees, particles were constantly being created and destroyed. Then, as the fireball expanded, it cooled and condensed into galaxies and stars. The discovery of the microwave background was strong evidence that this, most unlikely of stories of creation, was true, and it is widely accepted as the lingering remnant of the heat from that initial explosion. If quasars *are* among the most distant and powerful objects known, then it is possible that they represent a glimpse of the kind of body that existed in the early Universe, and we are seeing them by the light that left them nine thousand million years ago, shortly after the Universe was formed.

More tantalizing, and even more frightening, perhaps, than the birth of the Universe are the concepts of black holes and the death of stars. When a star has used up the nuclear fuel which sustains it, it begins to contract under the gravitational pull of the particles within it. Depending on its mass, this collapse may continue until, so the theory goes, the density of the star is so great that it becomes a black hole. The gravitational field of a black hole is so immense that nothing, not even light, can escape from it. As it cannot be directly observed, its existence must be inferred from the strange behaviour of adjacent bodies.

All these ideas are Martin Rees' everyday subject matter. He is Plumian Professor of Astronomy and Experimental Philosophy at the University of Cambridge, but not, as his title implies, an astronomer, using telescopes of one kind or another to gaze at the skies. Instead, like all cosmologists, he takes those observations and endeavours to make sense of them. It must, I thought, be an uncomfortable and intellectually exhausting business to bring the cosmos down to breakfast. How, I wondered, does Rees cope with it? Does he sit down and try to imagine a black hole, for example, or does he avoid the problem by reducing it all to mathematics? Does the subject matter have a philosophical or even religious dimension for him?

It was armed with such questions that I went to talk to Rees in his rooms at King's College. The wall panels are decorated by the Bloomsbury artist, Duncan Grant, and we began with the question of images. I think of galaxies as bright balls of light; how did Rees picture them?

———————

I think of them in more or less the same way. I would say that most of what I do involves physics which is not particularly exotic. In understanding galaxies one is just trying to apply the laws of physics that we understand fairly well on Earth to a large scale, unusual phenomenon. There are, however, some contexts where the physics that's relevant is very unusual, and I can mention two: one is in gravitational collapse and black holes—objects of that kind which we think are important in galactic nuclei. The other context concerns the early stages of the Universe soon after the Big Bang, when everything was squeezed to conditions where the physics is very uncertain, and is certainly very extreme.

'If we talk about black holes, or the beginning of the Universe, I'd just like to know what mental image you have there. Is it a mathematical image or is it a physical image?'

Whenever possible it's a physical image because I tend to think much more easily in terms of pictures and diagrams than in terms of equations. So whenever possible I try and have a physical image. And, in fact, if I compare what *I'm* trying to do with what the physicist is trying to do when he thinks about sub-nuclear particles, I would say that, in a sense, I have an easier task in visualizing my object of study. The objects I'm concerned with do obey ordinary, classical physics; we can think of them in terms of ordinary space and ordinary time. And so in a sense they're not so counter-intuitive as the concepts which are essential in trying to understand nuclear physics.

'When you think of a black hole, do you actually think of a black hole?'

Yes. I think of what it would be like if you were sitting just outside it; how light rays would bend, and what the forces would be; how it affects magnetic fields, and how it affects matter orbiting near it. So I try and think of this in a fairly straightforward way. Obviously one has to do some mathematics in order to ensure that one's picture is correct, but wherever possible I like to draw diagrams and like to visualize what would be happening.

'Let's take the other image then, near the beginning of the Universe. What worries me is that you can set a date, which is something like 10 000 million years ago, for the origin of the Universe, and what I find absolutely impossible to think about is going back even one year beyond that point.'

The idea that time has a beginning and an end is hard to take. It's interesting psychologically that many people are unhappy with the idea of a beginning rather than the idea of an end. The idea that the Universe will end with a big crunch may be in a sense unappealing, but it's not a conceptual problem in the same way as the idea of there being a beginning to time. I think this is an instance of where commonsense notions are going to be transcended under extreme conditions, and in the Big Bang we will find the conditions were so extreme that the idea of a direction of time probably can no longer be justified.

'Does that have implications for the way you like to think about problems?'

I think when one gets to those extremes, then one can't think in pictures as I like to do. That's why most of my work, perhaps, has not been in those areas because then one does have to think in terms of mathematical concepts, and not try and concretize them in any way.

'How did you come to choose cosmology?'

I started out as a mathematician and came into this subject in the mid-1960s as a research student. That happened to be a particularly exciting time because it's when the quasars were discovered and also the so-called microwave background radiation. This was the first evidence that there was some sort of radiation filling space that was a relic of the Big Bang, and it then became possible to talk about what the Universe was like in its distant past, and to think about problems involving relativity and rather exotic physics, without feeling that one was going to be cut off from the observations. So I really came into the subject because there were these new developments. And it was in a sense a good time for a young person, because if the subject's entirely new then the experience of older people is at a heavy discount. I was lucky in that sense to be introduced to this subject at that time, because I'd not been the kind of person who'd been keen on astronomy from earliest youth; I just came into it through an almost arbitrary choice.

'You're a theoretician, and yet the thing is there aren't too many facts around. Does that present a problem, and how do you get access to the experimental data?'

Well there are very few facts relating to cosmology proper, the nature of the Universe as a single entity but, of course, there is a great deal of data relating to individual objects in the Universe—individual galaxies and individual stars in those galaxies. And I am the kind of theorist who is particularly interested in specific data and particular objects, so I do follow the literature and talk to my colleagues and even suggest observations. And my work is therefore of a rather synthetic kind where I try and put together data obtained by different techniques, to see if one can understand it physically. Of course there are some theorists whose work is of a more mathematical and more deductive kind. People who do work in relativity and people who work in particle physics tend to work in a deductive way. They write down some equations and then follow through a long chain of argument to solve them. I tend not to do that.

'So what kind of data do you work with?'

It may be observations made with large telescopes of the spectra of a distant quasar; it may be results from a spacecraft looking for cosmic X-rays which tell us about hot gas in intergalactic space; it may be the results of radio observations which tell us the regions in space from which radio emission is coming. And by relating this body of data, I try to develop a model for a particular object under consideration. In particular I'm concerned with the nuclei of galaxies where it seems that some kind of runaway catastrophe has happened which causes something in the centre of the galaxy to put out far more energy than ordinary stars. The physics here is not primarily nuclear physics as it is in stars; it's almost certainly relativity where black holes form and more extreme conditions are involved. So what I try to do is to use the physics that I think I understand to see if one can develop a description of these objects, and hope that this description fits the facts which I have available, and fits also what new facts come in over the next few years.

'Does that mean to say each morning you get up and work on the model? Are you like a sculptor who has a theory there, and each morning you come in and model it?'

Well in a sense, in that I will think about a model, and then I'll discover some new observation I hadn't heard about before that discredits it; or I'll find that some new fact fits in with what I already thought, and that may suggest some particular calculation which is worth doing. What one starts off with is a general picture—what one might call a 'scenario'—of what's going on, and in terms of this scenario tries, on the one hand, to interpret the observations, and on the other hand to think of detailed bits of physics; because, of course, it's too difficult to solve the whole problem in one go. What we have to do is to decide which parts of physics are relevant and to work on bits of the calculation in the hope that the whole thing comes into focus.

'But when you say a calculation, do you mean a morning's arithmetic, or do you mean a really quite long and involved thing that could take several months or so?'

It could be. In fact the calculations I do, tend not to be of that kind. The problem really is in deciding what is the relevant calculation to do, and in deciding what bits of physics are crucial. That's where some kind of judgement comes in insofar as it comes into the subject at all. One has to decide what are the important facts, and then see if one can reduce the number of possible theories. To start off with, of course, we have just a few facts, and it's easy to think of a dozen possible interpretations. As more data accumulate, then some of these ideas fall by the wayside, but one hopes that at least one of them will turn out to be the basis of the right detail model. I don't claim to be very successful in this. I've certainly wasted a tremendous amount of time developing ideas which have turned out to be completely along the wrong lines in the light of subsequent data.

'When you say a lot of time . . .'
Probably half of my time,

'Is that really totally wasted work?'
It isn't wasted because there are genuinely different approaches one can adopt to a subject. It seems to me that if we have a phenomenon where there are, to start with, various alternative interpretations, then the right way to proceed is to follow up the consequences of the various alternatives in the hope that you'll discover a theoretical inconsistency in

some of them, or you'll discover new ways of testing them. And then you hope that by, as it were, running the horses against each other some will fall by the wayside and you will end up with an obvious front runner on which you will then focus your attentions. Here there's an interesting psychological difference I've found among different scientists. Some people aren't motivated to work on any theory unless they believe at the time it's almost certainly right. They have the attitude of a sort of advocate: they've got to have their theory and defend it against all criticisms. I tend not to be like that. I quite happily explore the consequences of possibly inconsistent alternatives, and I would argue that that's quite a good way in which to decide which theory is most likely to be correct. In a sense my strategy's a low-risk one in that I am never in the situation of having devoted a tremendous amount of effort to a single theory which is then discredited.

'Is that a little cowardly, or do you think it's genuinely the right tactic?'

Well I wouldn't say it's the *right* tactic. It's the one that suits me best, because I just find it interesting to explore a lot of different ideas rather than focus on a single one. And I think it probably is a fruitful approach to adopt in the early stages of a subject because it's obviously most unlikely you'll hit on the correct answer right from the start. And it's not worth at that stage developing any single alternative in detail. But eventually, if you follow a particular topic, then you find that it's fairly obvious what the broad outlines are. It's then that one should start thinking about details. One could, perhaps, say more generally that if you follow a particular science then it needs different personality types at different stages in its development. You know, it's often said that Lord Rutherford would not have been happy doing particle physics today; he'd probably have done biology or something like that. And similarly, one can see that the style of thinking which is appropriate in the early speculative stage of a subject is no longer so appropriate when one has got to the stage of working out the fine details. I would hope that in the next decade the theoretical basis of cosmology will become firmed up, getting to the stage when it becomes no different, really, from other branches of physical science.

'You say that there isn't an awful lot of data. Does this mean that cosmology is dominated by giants? In other words, are there a lot of very good theorists who dominate the field, but rather few people, in contrast to other sciences, burrowing away like mad at small problems?'

The total number of people in our subject is rather small compared to other subjects. If one compares cosmology and the whole of astrophysics with any sub-field in physics, you'll find that in the kind of work I do the ratio of people to problems is extremely low. This has two effects. It means first that there's not a tremendous amount of duplication. You don't find that everyone jumps on a new idea and it's worked to death within a year, and that's a pleasant feature of the subject. But, on the other hand, you sometimes find that there just aren't enough people to do the nice calculations one would like to see done. There are lots of obvious calculations that one would like to see carried out in detail, and the number of active people in our subject is not sufficient to do it. That's partly because of the small number of people, and partly because in the last decade the subject has developed. As the frontiers have advanced their periphery has greatly extended, so there are more new problems to work on now.

'Is it quite a cohesive field then, where everybody knows each other and you all interact rather happily?'

With surprisingly few exceptions I would say yes. What I find so appealing about this subject is it has not yet got to the stage where one has to over-specialize. It's possible for a single person to follow most of the literature over a fairly broad range of all the relevant observations, and the relevant theory as well.

'It must be a delight to work in a field like that.'

It's a good deal better than the more highly developed fields!

'If there are few people that means there's not much money in the field. Even so, we're always complaining that the astronomers are very expensive; too much is spent on space probes and telescopes and so forth. Aren't the astronomers a bit self-indulgent?'

Well certainly astronomy, and particularly space research, is what would be called 'big science' in the sense that it depends very much on large pieces of equipment of a kind that no single laboratory, and sometimes no single country, can afford. So in that respect it's like high energy particle physics where we have these big accelerators in Geneva, and at Fermi Lab in the United States, which serve whole continents. It's certainly, therefore, true that the cost per active practitioner is very high and, obviously, when one is concerned with big sums of money, then the justification has to come from feeling that it's part of some general endeavour, and that astronomy is not done simply for the benefit of its practitioners. And I think you can perhaps argue that astronomy is a subject that *does* have broader impacts. It is, after all, a subject which can be explained at a general level, and in which there is enormous public interest. The most spectacular examples of this, obviously, were the planetary probes—the flybys of Jupiter, and things like—that which interested an enormous public in space exploration.

'But then it would also be true to say that movies interest the general public—and they're cheaper.'

Well are they? Let's look at the total takings of popular movies. They run into hundreds of millions of dollars per year. Some science fiction movies have individually grossed $100 million. And I'm sure if one asked the people who enjoyed those movies whether they would be happy to see a small percentage of the takings spent on space exploration, they'd be very happy to think it was being done. So I think that astronomers are in this respect, perhaps, in a stronger position than the high energy particle physicists whose subject bears little popular appeal. It's much harder for them to convey at a general level what's going on. But having said that, I don't want to be in any sense knocking the particle physicists because one thing which has become clear in the last few years is the strong interrelation between these two, at first sight, very different branches of fundamental science: the study of the very large scale—the cosmos—on the one hand, and the study of the sub-nuclear world on the other. It's becoming clear that in a sense the cosmos provides the only laboratory where sufficiently extreme conditions are ever achieved to test new ideas on particle physics. The energies in the Big Bang were far higher than we can ever achieve on Earth. So by looking at evidence for the Big Bang, and by studying things like neutron stars, we are in effect learning something about fundamental physics. And, conversely, if we want to understand the Big Bang, we now find that we are held up by not knowing enough particle physics. Whereas in ordinary stars the physics is the physics of ordinary atoms and ordinary nuclei, answers to questions about why the Big Bang

happened, and what it was like in the early stages, and why it contains
the embryonic galaxies, depend on progress in particle physics. So I
think one can say that there is a growing interrelation between particle
physics and cosmology—they are more dependent on each other—but,
of course, they are both sciences which do depend on very large
centralized pieces of equipment.

'As a cosmologist, it seems to me that one has either to have a
tremendous intellectual arrogance or a tremendous intellectual inferior-
ity. I just wonder how you cope with the enormity of the problems . . .
or does it not impinge upon you in your day to day thinking?'

Well I think the most striking aspect of cosmology that must be apparent
to any practitioner, is that one has been able to make any progress at all.
It's rather amazing that the laws of physics, which we can understand
from experiments on Earth and which apply locally, do seem to apply
back in the early Universe, and that we have been able to at least get a
self-consistent picture of the early Universe by applying the physics we
know. Of course, self-consistency doesn't guarantee truth, but we
haven't run up against any obvious contradictions. So it is surprising that
we are able to make any progress at all in grasping the overall structure
of the cosmos just based on the physics we understand in the lab.

'You say it is amazing and yet at the same time there's a funny
contradiction because you make it all sound so reasonable. Here you are
dealing with the enormity of the cosmos, yet you make it sound as if it's
rather like a bit of Meccano or a chemical reaction going on in the
laboratory. I just wonder whether you remain in awe of it?'

Well, one remains in awe of its scale; but, of course, complexity and
scale aren't the same thing. Indeed, they're the opposite. Let me give you
an example. We understand more about the centre of the Sun than we
do about the centre of the Earth, and that's because in the centre of the
Sun conditions are more extreme; all complex chemicals are broken
down. And similarly, one might argue that the early stages of the Big
Bang, where everything is broken down into the simplest sub-nuclear
particles, is a state where everything is simpler and smoother. So perhaps
the understanding of the Big Bang is by no means the most difficult
problem facing science. In one sense it's a really grand problem, but in
another sense it's perhaps a more straightforward problem where we

have greater hope of getting a definite answer than in understanding greater complexities nearer to home. And certainly I would say it's a far easier problem than anything in the biological world.

'I'm amazed that you say that, but I begin to understand what you mean in the sense that you're applying quite well-founded principles of physics. The fact that things are on a large scale shouldn't in any way fill one with particular awe or fear or anything else.'

Well, it should fill one with awe in a sense, but on the other hand, it should not make one feel too daunted and too depressed about the prospect of ever understanding it.

'Now you've given me a completely different image of cosmology to that which I had before. You say it's really not that complicated. If I were to ask you whether it has religious implications, our understanding of the cosmos, I wonder what your reply would be?'

I would say that it has no special religious implications compared to any implications that might have been drawn by an eighteenth century scientist. The eighteenth century scientist could not extend the chain of causality back beyond the existence of the solar system. We can go back a lot further, but in principle there's no difference: there's a certain set of things one can explain, and then we get to some sort of barrier, some kind of frontier which is the limit of our present understanding. I wouldn't have thought there were any particular theological or philosophical implications of modern cosmology, or indeed of modern physics. There are more likely to be such implications, I would have thought, in *your* area. However, one thing one does learn, I think, from the practice of physics is that even the simplest things are pretty hard to understand. Even the hydrogen atom is hard enough. And also one must be wary of extrapolating common sense notions: we know that the idea of space and time and ordinary stuff we see around us isn't a valid way of describing the sub-nuclear world nor, indeed, is it valid in describing the whole cosmos; we have to introduce new ideas rather far from common sense. And this ought to make one slightly dubious about any claim that we have more than an incomplete metaphorical understanding of any deeper reality.

'Do you think that cosmologists then have a more open mind? Do you, for example, have a more open mind about the fringe of science than other scientists, because of your experience in cosmology?'

Well, I hope I'm not completely impartial between sense and nonsense, but it could be that physicists are slightly less inclined to be dogmatically reductionist than many biologists now are. Perhaps it's fair to say, as some people have said, that the biologists now have the supreme confidence which one associates with nineteenth century physicists rather than twentieth century ones. I'm sure there's some psychological truth in that, but I wouldn't go beyond that to say that one should be ready to invoke any fundamentally new principles in biology at the moment. I think one should keep a slightly open mind to the possibility that there may be some fundamentally new feature that emerges when you get above a certain level of complexity. We have no firm reason for disproving this, although I agree with most biologists that it's not helpful to devote much time to these ideas now. One ought to pursue any science, it seems to me, on the basis of the simplest principles, until some obvious glaring contradiction emerges. And if we can't understand some phenomenon then that simply means that we have not thought long enough or hard enough about it. It's not a reason for throwing in the sponge and saying it involves some fundamentally new science. I can give examples of this from within my own subjects. Some people have argued that the properties of quasars and galactic nuclei are so extreme that we already are transcending the bounds of conventional physics—that we need some fundamentally new laws of nature to explain them. It seems to me that this is a premature throwing in of the sponge because we can think of many previous instances where a phenomenon could not at first be accounted for in terms of ordinary physics. Superconductivity, for example, was not explained until about 50 years after the basic laws of quantum theory. And there are many phenomena in astronomy like the sunspot cycle which we can't properly explain even though we fully believe they're just ordinary gas dynamics. So one ought to accept that even if we understand the basic laws and can write down the equations, the complexities of the solutions to those equations are such that we will perhaps never exhaust them, and so perhaps will never have any firm grounds for believing that they are inadequate. But I think one should be slightly open-minded about the possibility that they may in some sense be inadequate.

'Do you think that there will be some things that remain unknowable as far as cosmology is concerned?'

Quite possibly. There's a great debate among physicists about whether it will be possible to develop a unified theory of all the forces. It could be that we *never* develop such a theory. On the other hand, if we do, it could be that such a theory will give us a deeper understanding of, for instance, the Big Bang in the sense of showing that there's a certain unique self-consistency about the way the Universe is. Now even if we do have that complete understanding, then it's an important achievement in science, but it's not, of course, the end of all creative science. People sometimes use the analogy of the chess game. The physicist is like someone who's watching people playing chess and, after watching a few games, he may have worked out what the moves in the game are. But understanding the rules is just a trivial preliminary on the long route from being a novice to being a grand master. So even if we understand all the laws of physics, then exploring their consequences in the everyday world where complex structures can exist is a far more daunting task, and that's an inexhaustible one I'm sure.

MICHAEL BERRY *was born in 1941 and is*
Professor of Physics at
Bristol University

CHAPTER 4

THE ELECTRON AT THE END OF THE UNIVERSE

MICHAEL BERRY
Theoretical physicist

PHYSICS is regarded as the hard edge of science. For most non-physicists the paradigm is still that of Newtonian physics, operating in a Universe whose every motion is theoretically predictable according to known laws. Yet in reality, an essential feature of modern physics is *uncertainty*. In 1927 Heisenberg deduced his celebrated uncertainty principle—that it is inherently impossible to measure exactly and simultaneously both the position and the momentum of a quantum particle. This has profound implications for the behaviour of matter at a sub-atomic level and effectively destroys the notion of a simple deterministic account of the universe. Nevertheless, outside this exotic realm of quantum mechanics, we still think of the physical world as an orderly and predictable place. It comes as something of a shock, therefore, to learn that a part of modern physics is preoccupied with the chaos of everyday situations. A major topic of study, for example, is the chaotic behaviour which occurs when a fluid pouring down a channel or pipe changes from a smooth, orderly pattern of flow to a turbulent one, characterized by swirls and eddies whose fluctuations are irregular and unpredictable. Chaos is mathematically respectable but, to the non-physicist, as difficult a concept as that of infinity. It makes one wonder whether physics is now losing that reliability and predictability which was always so reassuring.

Michael Berry, who is Professor of Physics at the University of Bristol, works on wave problems. It is a field which is concerned with many physical phenomena, ranging from the behaviour of radio waves and light to the motions of sub-atomic particles. Berry has addressed problems as diverse as the twinkling of stars and how it is that male moths can follow the trail of a female pheromone as it diffuses over long distances, perturbed by wind currents. His major tool is mathematics, which has always been the means by which physicists encode and understand the world. Newton's second law of motion, for example, which defines the force applied to an object in terms of its mass and acceleration, is expressed as $F = ma$. If known values are substituted for any two of the terms, the equation can be used to discover the third in an infinite number of situations. Michael Berry applies some of the newest ideas in mathematics to wave problems, and it was by using a type of mathematics known as catastrophe theory that he found it possible to describe the complexity of patterns made by sunlight shining down on to the bottom of the swimming pool. I had always thought that every school-child knew—and had known for several centuries—about the refraction of light, and the way it changes direction at the boundary between, say, air and water. What, I wanted to ask him, has become of physics? What has happened to all these laws and equations that once seemed to have the Universe sewn up?

———

In the past, a theoretical physicist who studied waves would have been conceived of as a person who found the solution to certain types of equation. The equations were well known, long known, and very simple to write down. There are certain standard mathematical representations of a wave which is travelling in empty space and which has come from a point source, or of a wave which has plane wavefronts, so that the rays of energy travel in straight lines and all the different rays are parallel to one another. Formulae of this kind tell what happens if a wave hits a·diffraction grafting or passes through a lens. But all of those solutions have the property that they are possible only because of some simplifying circumstances in the physical situation—the cylindrical symmetry of a lens for example. Now it's been realized recently that in what's called the generic case—the case that has nothing special about it, and which people previously thought they could say nothing about without big computers to solve the equations—that this, actually, is the case to which a wonderful new regime applies. There are universal forms which emerge in these generic cases. An example of a waves problem which is a generic case is the refraction of sunlight by the wavy water of a swimming pool, where you see the bright lines of refracted light focused on the bottom of the pool. If you had asked an optical scientist ten years ago 'How do you explain those lines? I want to know what I actually see down there', then he could have said one of two things. Probably he would have begun by saying 'Oh! That's a rather trivial problem, it's just refraction of light, that's Snell's law, Snell sixteen hundred and something. We all understand the law of refraction'. But if you'd pressed him and said 'No, no, that's not enough, I want to know how the law of refraction gives rise to those morphologies that I see on the bottom there.' 'Ah', he would say, 'that's very difficult—you need a computer for that. But then it's trivial. With a big enough computer you can work out what those patterns are.' A computer would, indeed, provide you with a simulation, and you would, indeed, find that with quite simple patterns of waves on the water surface you would get quite realistic looking patterns of focused lines on the bottom of the pool. But still you would be missing the understanding. You wouldn't be able to answer the question 'Why is it I always see, for example, junctions of lines of this sort, and not that sort?' He wouldn't have been able to answer that. Now one *can* answer it because precisely this kind of morphology is classified by catastrophe theory. This means that one has a sort of library—a very small library— of universal forms out of which such short-wave patterns come. They are short-wave patterns because the waves of light are very, very small compared with waves of water. And this is a very intensely developing subject. Now one can really hit the heart of what actually one sees with one's eyes, which one previously was unable to do.

'Is this a new kind of physics that you're dealing with, or does it fall within conventional physics?''

I would characterize it in the following way. The image of a physicist is often of somebody who's seeking to discover the laws of Nature. Now, as it happens, it does appear that those laws are mathematical and so one speaks of trying to discover the fundamental equations for elementary particles, or fundamental fields that they satisfy, and so on. That activity's going on and, indeed, is undergoing something of a revolution now. It's a very exciting time. But what's being realized now is that concealed in the old-fashioned equations, long known, that describe matter on more familiar scales, there are new solutions, new phenomena which can only be brought out by using modern kinds of mathematics. And that's how I would characterize the feeling of excitement that is occasioned by the solution of ancient problems of this kind. One of the most difficult—it's still resisting solution, and I would characterize it as being the most important problem in theoretical physics apart from the elementary particles—is understanding the problem of fluid turbulence. Why is fluid motion so often unpredictable? That's precisely a problem of understanding the solutions of the fluid equations which were worked out by Navier and Stokes 150 years ago. So it's certainly old physics from a particle physicist's point of view, but the problem is understanding how these equations contain in them chaos. It's very easy to find solutions which aren't chaotic, but they're not the ones you observe very often. If the viscosity is small enough as it is, for example, with water and air, then one gets spontaneously developing chaotic motions. You can't, in fact, use catastrophe theory to solve that problem. It's the wrong sort of mathematics, and these are more difficult problems and they're being very very intensely worked on. But they still fit into this category of simple equations having complicated solutions. In a way that's a very satisfying answer to the problems which people who aren't scientists often bring up. They say 'Here you work with these few equations. You can write down on one sheet of paper all the equations of theoretical physics, but I see the world as a rich and complicated and beautiful place. Aren't you brutally truncating it in that way?' The answer is actually that simple equations have complicated solutions; it's a very, very compact encoding. Now one is able—with the aid of mathematics—to make very substantial steps in a whole new class of decodings.

'Chaos—that's not a concept that I would have associated with physics. My image of physics is that it gives us a highly ordered image of the world or of the universe. Does this concept of chaos imply that there's no predictability?'

Sometimes, strangely enough, it implies just that. We, as theoretical physicists, were brought up to believe that fundamental chaos only entered with quantum mechanics—you had the indeterminacy principle, and so on—and that before the advent of quantum mechanics the universe was predictable in the sense that you had for example, Newton's laws of motion, which, even as modified by Einstein, tell you that if you know the initial state of the universe—all the particles and their positions and velocities—you can predict for ever more its behaviour. We believed that. We swallowed that particular myth although it flies flagrantly in the face of anybody who has ever used a pinball machine. But in fact what's realized now is that unpredictability is very common, it's not just some special case. It's very common for dynamical systems to exhibit extreme unpredictability, in the sense that you can have perfectly definite equations, but the solutions can be unpredictable to a degree that makes it quite unreasonable to use the formal causality built into the equations as the basis for any intelligent philosophy of prediction. I give you this example. Suppose you've got lots of colliding particles. You can think of them as molecules of oxygen, let's say, in a gas and they're in a box. You believe that they obey Newtonian mechanics. They don't quite, but let's suppose they do. And you measure their initial position and velocity precisely. Of course, one couldn't do that, but suppose one can. Then one could predict their motion for all time. But wait a minute. You can only do that if the system is completely isolated and so you say, 'Isolate it as best you can according to the laws of physics as we now know them.' Well there's one force that you can't screen out, and that's gravity. So unless you know the position of every single external particle in the universe which would have a gravitational effect on your molecules, you couldn't predict the motion. So let's estimate the uncertainty that arises from this source by considering the gravitational effect of an electron at the observable limit of the universe. You agree it's not possible to think of a smaller perturbation than that.

'A single electron?'

A single electron. Just one. There it is at the observable limit of the universe, say ten thousand million light years away. It has its gravitational effect, but you don't know where it is *exactly*, so that's the uncertainty. Well, you ask, after how many collisions will the little uncertainty that's produced in a motion by that electron be amplified to the degree where you've lost all predictability, in the sense that you make an error in predicting the angle at which a particular molecule will emerge from a collision by say 90°. You could reasonably say you've lost predictability then. Well, the amazing thing is that the number of collisions is only about 50 or so, which is of course over in a tiny, tiny fraction of a microsecond. That means that it's really unreasonable in a large class of systems to consider Newtonian mechanics as being predictable. The more realistic case, actually, is if you think of the particles as billiard balls on a billiard table—an ideal, perfectly flat, perfectly smooth billiard table—and this time you're considering the uncertainties as being produced by the gravitational force of people moving about near the billiard table. There always are such people milling about. You want to know after how many collisions of the billiard balls will their motion be uncertain because of this. And the answer is six or seven, and that's why no billiard player, even the best in the world, can ever plan a shot which would have even three or four consecutive collisions and have some reasonable expectation that he'll be able to successfully carry it out.

'I always had the image of physics as an exact science. Do you think this undermines physics as an exact science?'

No, I don't think it undermines physics as an exact science because generally, when one can't predict something accurately, as in these cases, one then finds that actually one didn't really want to. It's statistical properties that one really is more interested in. But it has implications for other sciences which would seek to use physics as their model. I'm thinking particularly of economics. I don't know anything about economics but there are physicists who do, and they're now very intensely studying the possibility that the models that economists use to predict the future from the present are equations of this sort. They may be very pleased to get exact equations or accurate equations and they might think that therefore, like physicists, they can now predict the future. But it's likely to be the case that the equations are, in fact, of the unstable kind. They might find that knowing the laws doesn't enable them to predict the future. That's something which is just slowly filtering into other sciences which would seek to use mathematics. In other words, they have an outmoded paradigm of physics.

'How do you actually go about your work? Do you collaborate or do you work in isolation?'

A bit of each. Sometimes problems come rather naturally and internally as developments from what I've done before. Many of these morphological problems in waves are like that, but sometimes they arise quite randomly as a result of conversations I've had with people. The one about the male moth and the female moth arose from a conversation I had with a biologist and then very remarkably it turned out to have all sorts of links with things I'd done before. It's really a mixture of keeping my eyes and ears open and developing themes, if you like. It's themes I think of basically. Many practising theorists of my type—that is those who do a range of problems rather than one particular thing—develop a *style*. This means that they're alert to problems of particular sorts and look at them in particular ways. I'm aware that this is not at all the conventional notion of how scientists work. I tend to think of myself not so much as finding things out, although I do that, as of *making* things. I think of myself more as a carpenter than as somebody who is given a problem, solves it and then moves on to the next one.

'Could you explain what "making things" means?'

Yes, theories. The things are theories. I mean when you make a mathematical theory of some phenomenon, there's a large element of hewing it into shape. You polish a bit here and there and you chop off a bit here and there and then you think 'Well there's a bit here that needs refining' and so on. The resulting production has two aspects. It has an internal aspect, which is its coherence, whether one's pleased with its structure and so on, and the external aspect, which is its relation to the problem that stimulated it and to which, in the happiest cases, it can give something back in the sense of predicting something or explaining something. There's very much this element of making this sort of aesthetic and creative judgement which I guess isn't terribly widely appreciated.

'Now when you talk about hewing, are you hewing from mathematics in the first instance? Is it a mathematical chopping and changing, finding the right bit of mathematics to fit the phenomena that you're interested in?'

Yes, but it's worth saying that this is a very different kind of mathematics than the mathematics which mathematicians create. They're creating theorems. I think of those as raw materials. I use them to create theories.

That's really quite different, and there's a considerable difference, sometimes almost incompatibility, between these two modes of thought. I'm very much *not* a mathematician—I haven't studied mathematics in a formal way. I've taught myself, so much of what I write would, I guess, be quite horrific to some mathematicians. A mathematician is concerned with the logical rigour of every step, but as the physicist Richard Feynman said 'A great many more things are known than can be proved.' And while a physicist wants to be right, he doesn't want rigour to turn into rigor mortis. And if you're trying to create some rather elaborate edifice, and calling on lots of branches of mathematics, you just can't afford to be perfectly rigorous with every step. It's as though, for example, a printer would, in printing a book, have to insist that every letter was absolutely perfectly formed, that there wasn't a shadow of a smudge anywhere. You'd never get beyond the first line if you did that. That's really what I mean.

'So it's understanding via mathematics?'

Exactly so. Modelling and understanding.

'And you'll take any bit of mathematics from anywhere and use it in a way that will give you insight into your physical phenomena?'

That's right, and it's here that there's been quite a change in my and many other physicists' attitude to mathematics in the last decade. We took the view previously that modern mathematics was certainly logically very exciting and interesting and worthwhile under the terms of mathematicians' work, but as far as we physicists are concerned we knew all the mathematics we needed to know and all this modern stuff was of no assistance to us. Well I've changed my view totally on that. I now take the opposite view, that there's no piece of worthwhile mathematics that has been—or will be—invented which cannot and will not some day be of use in describing some aspect of the universe. I'm forced to this view because I've found it to be so. I've found the most surprising pieces of mathematics—topology and number theory, catastrophe theory and so on—to be just what I've needed for particular physical problems. That's the practical reason why I'm forced to this view. But even theoretically, when you think about it, it's fairly obvious. After all the whole universe, which it's the object of science to describe, is more complicated than the inside of one's head and that's where mathematics comes from. That's my rationalization for this change of view that I've had over the last few years.

'You see yourself as a theoretical physicist. What's your relationship to the experimental physicist?'

Very close, of course. Physics describes the real world. It isn't a sort of low level mathematics, which it would become if one lost contact. It's very important to always realize that there are phenomena, that there is a world outside our heads that we're trying to explain. Otherwise it's a curious game, a form of self indulgence which I think is intellectually not very worthwhile.

'But when do you interact with other physicists? When you're hewing your ideas, is that a very personal, isolated process?'

Again, this depends. I do sometimes work with other theoretical physicists, my research students, for example, and other colleagues. Indeed it's a valuable part of a research student's training. Any theoretical physicist has to learn how to get a theory into an aesthetically worthwhile shape. But mainly I work by myself.

'You keep talking about the aesthetics. Can you explain what you mean when you speak about the "elegance" of the theory?'

Very difficult. Taste is a difficult concept to define and it's something that's appreciated by people who know it and not by people who don't. I don't want to sound terribly aristocratic, I certainly don't feel that way. You can only explain it by analogy. A piece of music, let's say, can be ill-constructed or well-constructed, and you can after a while hear—learn to hear—the difference. It's like that with theories. I wouldn't like to give the impression that there could be no objective way of assessing the aesthetic value of a physical theory, but I don't know one.

'Is it something like when you hear a good bit of music, you get a nice feeling down your spine?'

Oh very much so. That happens, but that's part of the aesthetics. I don't call that objective. I mean it's objective in the sense that the tingling really happens and it's a psychically and, presumably, neurally real phenomenon. That's not what I meant. I meant that you could not communicate it to somebody who doesn't share the feeling. I don't think that's possible.

'To what extent, then, would you describe your choice of problems and your solutions to them as partly an exercise in self-indulgence? Doing things that please you enormously?'

When I said before that certain other kinds of activity were self-indulgent, I was implicitly defending myself against that charge on the grounds that theoretical physics of the kind that I and my colleagues do has some connection, however distant, with the real world. But there is, of course, a large element of play and that's very necessary to the successful development of theories. If you keep very narrowly in mind a particular practical goal, for example, you often won't solve the problem, at least at the level that we're speaking about. Some problems have been solved by this goal-oriented approach, but not the problems I've been speaking about. Of course one could say that that's only an excuse, and the fact that one's playing means that it's a form of self-indulgence. In defence against that I would only say that by that same criterion so is poetry, music, writing, and philosophy, and so has been much of previous science. I don't want to be dishonest and claim that the reason why I'm doing theoretical physics is because its productions are useful in industry. Occasionally they are, even the kinds of things that I do, but I don't want to claim that, because that would dishonestly represent my motivations.

'But when you say "play", do you mean having fun? What do you mean by play?'

Play means exploring every possible suggestion of an analogy or an allusion to some other part of mathematics. If one starts by solving one problem and suddenly finds an interesting by-way opening up, which looks as though it would be more fruitful and more rewarding, one isn't afraid to go along it. One does remember the original problem and come back to it, but one can spend a year sometimes going off and then finding that this unexpected pathway, this unexpected fork, takes us along a path which is very fruitful and surprising.

'Does it matter to you terribly whether what you do does have practical implications? Is that important to you?'

It's very pleasing when, as has happened a few times, something I've done has been useful. As I said, it would be dishonest of me to claim that that's my motivation for doing it. It isn't. But it's very nice when people from industry come and talk to me, and I'm always very happy to discuss the details of their problems and sometimes give minor technical help, such as solving equations and so on. But what is a much more difficult problem is the question of research which is used badly in my terminology. I'm thinking particularly of military research. There one always has to reckon with the possibility that what one does, or what one's colleagues do—which one is associated with as being a member of the whole enterprise—will be used in good and bad ways that one can't foresee. There's been a lot written about that, and a lot discussed on the question of responsibility of scientists for what they do, and my position is the following. In a complicated society such as ours, which is an interactive society, everyone's actions have effects which he or she can't foresee. I'm thinking, for example, of a man who makes steel in one of the steel works. He's certain that a fraction of the steel he makes, which, if he's mathematically minded, he can work out, will be used for purposes which he might not approve of. The steel will be used for making guns, knives and the like, and that's a mathematical certainty. Philosophically it's exactly the same as what we physicists do. We create something and then it becomes public property. What we make is part of our culture. We're making cultural artefacts, and we can have no control over what they do, after they've left us. But of course, that leads to the question of drawing lines and, in principle, it is impossible to draw a line which covers all circumstances. Since I do think that the application of science, especially of recent physics, to military purposes is a monstrous perversion of its aims and ideals, my personal line is not to do any work knowingly for the military or for immediate military ends. That's where I draw the line, but other people can draw lines differently.

'But it's very hard to predict what's going to happen in the future, so the same must be true in relation to the application of one's research work. For example, your work on how moths trail the scent of the female in wind currents obviously has military applications in connection with the spread of biological or deleterious agents. Doesn't almost anything one does have the chance of becoming used for, as it were, evil purposes?'

Yes, you're exactly right, and indeed the very example that you mention is well chosen. One of the papers I had to read in connection with this biological problem of the diffusion of pheromones was written in conjunction with somebody from Porton Down, and the interest that they have is obviously in what happens after the explosion of a chemical weapon. The distribution of death—and let's not put it any lesser way— the distribution of death will be governed by exactly this kind of mathematics. On the other hand, the distribution of pheromones is very important to the people interested in pest control who try to fool male moths with false scents and then catch them, and thereby save the crops. It's an immensely difficult problem. I think in every case, in every type of research that one does, one can rather easily and without too much imagination, find a military application. Every year the US Department of Defense publishes a list of aspects of fundamental physics on which it would like to see further research done. There isn't much left out. Modern warfare uses the whole range of science. In the battlefields, the oceans, the atmosphere, space, the deserts, the forest, everywhere you can think of there is an application for some kind of physics or another. I go back to what I said before, which was that if one wished to live a life which didn't in some way aid what I would call the forces of darkness, then I don't think one could do it in this society at all.

'That has the implication that you would not support the argument that has been used recently in relation to biology, that there are certain sorts of research that one shouldn't do in principle; that certain things should remain unknowable because of the danger of applying the knowledge.'

You're right. That isn't the view that I take. It's a very difficult question and one can clearly foresee such terrible things. Civilization is an uncertain enterprise and one of its uncertainties, which is also a tremendous adventure, is this intellectual journey into the unknown. I'm sorry to sound so melodramatic but that's what it is, and I think it won't work to place a moratorium on whole areas of research in this way. Although the mechanism would be extremely difficult, I could see and, indeed, would like to examine, the possibility that certain types of *applications* of research would be proscribed. Indeed, certain types of bacteriological warfare research were for a long time voluntarily not conducted by government, so it is possible. But I think one doesn't achieve that end by stopping whole areas of research.

PROFESSOR CHRISTOPHER ZEEMAN
*was born in 1925 and is Director of the
Mathematics Research Centre at
Warwick University.*

CHAPTER 5

PRIVATE GAMES

CHRISTOPHER ZEEMAN

Mathematician

MATHEMATICS is not arithmetic. Though mathematics may have arisen from the practices of counting and measuring it really deals with logical reasoning in which theorems—general and specific statements—can be deduced from the starting assumptions. It is, perhaps, the purest and most rigorous of intellectual activities, and is often thought of as queen of the sciences. To the outsider, however, it seems like a private game, the manipulation of symbols to uncertain and unwordly ends.

Many of the ideas in mathematics seem to lie beyond the bounds of common sense. It is obvious, for example, that the set of numbers 1, 2, 3 . . . is infinite and there cannot be a highest number. It is also reasonable that the sum of an infinite series such as $1 + \frac{1}{2} + \frac{1}{4} + \frac{1}{8} \cdots$ is finite and equal to 2. But much less easy to understand is that the infinity of whole numbers equals the infinity of fractions but is less than the infinity of decimal numbers like $1.1276 \ldots$ Even when the nature of the problem is accessible to the layperson, its relevance is often hard to grasp. Consider a famous mathematical problem—Fermat's last theorem. The problem is to prove that there are no whole numbers, other than $n = 2$, which satisfy the equation $x^n + y^n = z^n$ where x, y, and z can be any whole numbers greater than zero. Why should mathematicians devote so much effort to the proof?

For scientists and applied mathematicians, mathematics is, of course, an essential tool. They need the right pieces of mathematics to enable them to express and solve their problems. But for the pure mathematician, the interest seems to lie in mathematics itself, rather than in its application to the real world. It was to try and glimpse this fascination, and to understand why someone should wish to devote his life to such problems, that I wished to talk to Christopher Zeeman.

Christopher Zeeman has been Professor of Mathematics at the University of Warwick since 1964. His strong physical presence and evident pragmatism run counter to the rather ascetic image that such a position might suggest. Yet the core of his work is in pure mathematics, in what seems a rather abstruse branch of geometry—topology. This deals with the properties of shapes when they are knotted and stretched, not just in three dimensions, but in many. But he also likes to apply mathematics to a wide variety of problems. In particular he has earned a certain notoriety for his use of catastrophe theory, which classifies various kinds of unstable situations, to illuminate a number of biological problems, including the medical condition, anorexia nervosa. As we talked, snow began to whiten and obliterate the landscape beyond his office windows. It seemed a particularly appropriate setting in which to try and gain access to abstract and unfamiliar territory.

———

I was always very good at maths at school. I can remember a problem which I couldn't solve when I was seven, and my mother showing me how to solve it using x's. It was my first introduction to algebra. I was absolutely flabbergasted at this technique, and I've never forgotten it. And then at school maths was always very easy. When I went up to Cambridge I was thinking about reading nuclear physics because that was fashionable at the time. But I found applied maths at university got very 'messy' and I much preferred the pure. So I became a pure mathematician for twenty years. Even so I think my heart's always been really quite a bit towards applied, so I've been doing some of that recently.

'Can you remember what the problem was that your mother solved?'

Yes, sure. You had a room which was 3 yards by 4 yards, and you had a carpet which was half the area of the room and left an equal surround all the way round the wall. What was the shape of the carpet?

'Isn't it a little bit unusual for one's mother to solve a problem using x's? Was your mother mathematically inclined?'

Oh no, not a bit! I can't remember discussing maths with her ever again. But it's the earliest thing I can remember, and then at school I just found it all very easy.

'Do you think that all mathematicians find they are good at maths very early on?'

Yes. I think they all find it easy. They leap on to the concepts quite quickly.

'Could you try and define what mathematics is?'

Well, it's very difficult to define because it's a thing you only get used to when you've been doing it for a while. I mean, if you ask a child, 'what is arithmetic?' he won't know. But if you ask him to do sums he gradually gets used to doing sums and tables. Well, the more mathematics one does, the more one gets familiar with the subject, and the larger you realize its language becomes, and the broader the concepts. I can begin to say to you certain things about mathematics which you would recognize: numbers, techniques of infinity, and techniques of proof. But gradually you build up concepts upon concepts. First of all, from numbers you build up the concept of space and dimensions. Then you build up the concept of calculus and things moving in space and so on. Then you build up the concept of differential equations, trying to mimic the way things move, like planets round the sun, or populations growing in biology. And, finally, the whole language is about something quite different to the original numbers that you thought of.

'Do you think it requires special qualities to be a mathematician, in the same way, say, that it does to be a musician?'

Well, I don't know what those qualities are, but I do know that among students, the good ones are automatically good and the poor ones you can't really make better. They may love it but they reach a ceiling; they can't master it. Maybe it's something to do with the way the mind processes information or puts it around in certain patterns—I don't know.

'Now, you're a geometer. Are there lots of different sorts of mathematicians?'

I think there are probably three basic types: algebraists, analysts, and geometers. And this is a question of taste. And whatever your taste is, you generally prefer to do that quality of problem. Algebraists like systems, complicated systems. Geometers like to picture things. They like to reduce their equations to geometrical pictures in their minds. These are maybe two- or three-dimensional pictures, but they enable one to glimpse which theorems are true in higher dimensions. And analysts like the fine details of sewing the thing together in very careful bounds and accuracies and approximations and so forth.

I think they must be different psychological types because one of the things we do here at Warwick is to run big symposia. Each year we choose a subject and have maybe 40 or 50 visitors in that subject. My secretary arranges the housing for all these people, and she finds that the different years have completely different housing problems. The geometers and the topologists generally all bring their families and have lots of parties and often stay longer than they intended. The algebraists are very precise, often come alone, without their families, on precisely the day they said they would three years previously. And the analysts are totally unreliable. They say they'll bring their families and then they turn up with their mistresses, and they never come on the days they said they would. So they're completely different psychological types.

'Now, you decided you were going to be a pure mathematician because the applied seemed very "messy". What do you mean by "messy"?'

Well, I'm very much a geometer at heart and geometry is very clean. You're given some circles and lines or ellipses in some configuration or other, and you've got to prove that they have some symmetry, or intersect in some other surprising and beautiful way. The proofs are rigorous and aesthetically completely satisfying. But in applied maths you have to fit data, maybe, or because you can't solve the problem precisely, you have to expand in some ghastly series which may or may not converge . . . and everything gets very shady towards infinity.

'And you don't like approximations?'

Well, I begin to now! When I do applied maths I *do* approximate things. When I do calculations, I do very crude calculations in my head, which I much prefer to using a computer, because then it helps me to get the gist of the idea. I thoroughly believe in that. But when it comes down to proofs, well there's a quality of messiness and imprecision in the proofs that is distasteful. Certainly the type of applied maths that was fed to us as undergraduates had this messy quality especially when it invaded the proof area. And of course in those days one had to do 50 per cent pure maths and 50 per cent applied, which was mostly theoretical physics. And although I came top in applied, my heart was in the pure. And I went on to do a Ph.D. in Pure Mathematics under Shaun Wylie.

'Now how did you choose your Ph.D. problem? What sort of problems do mathematicians set themselves?'

Well, I was very intrigued by knots. And I was particularly intrigued by the algebraical machinery you need to actually prove knots exist. And I tried to prove you could tie a sphere in a knot in five dimensions.

'But why knots? Can you remember?'

Yes. I first went to Wylie's courses on topology because I thought it was a funny sounding word. T–O–P–O–L–O–G–Y. I couldn't think what it meant, you see? So, I just turned up at the lecture to find out what it was. And then, to my surprise, I found it was about the most geometrical subject that we were doing.

'You really didn't know what topology meant?'

Sure I didn't. Anyway, that was when I was in about my third year as an undergraduate, I guess. And I was very intrigued because the subject was so geometrical, and incredibly elegant. Also, Wylie was a very nice person. So, given the combination of a very elegant and geometrical subject, and a nice man, I asked if I could be his student.

'And your problem was tying knots in a sphere in five dimensions?'

Yes, but because these things are all interrelated, my thesis was actually very algebraical. I set up this great *algebraical* machinery so that I could solve a *geometrical* problem. And, in a way, that was a bad thing, because I got stuck in algebra for about six years when I was, at heart, a geometer. But after that I switched back to geometric topology, away from the algebra, and I felt much happier. I proved many theorems then. That was my first big change of subject.

'Was your algebraic topology successful? Would you regard that as important work?'

Well, it was modest. I wouldn't say it was world-shaking. My reputation as a mathematician rests on my work in geometric topology in the following eight to ten years.

'And what was your problem there?'

Well, again, it was this knot theory problem. What I had tried to do in my thesis was to prove this wretched sphere in five dimensions was knotted. And I'd invented this elaborate algebraic machinery that, in the end, failed to prove the thing was knotted although it did turn out to be quite useful for many other problems in topology. Then seven years later I sat down one Saturday morning—I had a free Saturday morning—and I thought 'Well, I'll have another crack at this darn problem.' And lo and behold, while I was trying to analyse why my previous attempts had never seemed to work I suddenly found, to my surprise, that I had proved the opposite: that you could *untie* the knots in spheres in five dimensions—but using geometry! You see, to prove that you can *do* something like untie a knot you have to use geometry. But to prove that you *can't* do something, that it always remained tied, you have to use algebra. That's the subtlety of this subject. Anyway, I used the geometry and I untied the wretched thing. And I was so excited that I spent the whole weekend writing this paper up, about 20 pages. And then late that night I confess I went and sat on the lavatory and while I was there the real flash of inspiration struck me like a bomb. I suddenly saw how to reduce the proof from 20 pages to 10 lines. Then I immediately saw how to generalize the whole thing to higher dimensions, how to untie 3-dimensional spheres in 6 dimensions, and 100-dimensional spheres in 103 dimensions. This led to the solution of a whole cascade of problems, that I had not previously anticipated. The first step was really the key.

'So it really was a 'Eureka' situation?'

Hmm . . . I suppose, yes, that's true. One often gets these flashes . . . no, *not* often! One rarely gets these flashes of insight where you get a really new discovery, but when you do, you can often pinpoint them. Many great mathematicians in the past have noted that. Poincaré wrote a book, *Science and Hypothesis*, where he lays his finger on the moments when he first thought of certain things.

'You don't know why it was that weekend?'

No, it wasn't a particularly different weekend.

'Well, it was obviously a very important weekend in your life. If we take the time leading up to it, you had spent six years not solving the problem. Doing what you thought was competent mathematics, but not solving the problem you had set out to solve.'

One very seldom does. Often one doesn't solve the original problem. That was a little problem you see, the one I set out to solve. The algebraical stuff that I developed was quite solid, good solid stuff, much bigger. But eventually solving the little problem seven years later was the key insight to a whole new development. After I'd solved that little problem I was able to settle down and work hard and use the techniques which had come in that flash of insight and develop a whole theory. And I subsequently had about a dozen research students who got Ph.D. theses out of that line of work and it led to the development of many new theorems in geometric topology.

'What does it mean to work hard? I don't have the feeling of what it means to work hard at mathematics. You get up in the morning and then what . . . ?'

If I'm really determined to tackle some problem it's possible I might spend as much as 18 hours in one day where I just sit at my desk or stand at the blackboard. Or go from the desk to the blackboard, operate between the two. And I just concentrate on this one problem, struggle to get a notation going. Sometimes if I've got too much in my head I have to go for a walk and just think about the problem all the time I'm walking along. It's a problem of communication between the hand and the eye and the brain. I have to write and watch the pattern emerging. The notation is so complicated I can't possibly keep it all in my mind so I have to write it down. And then once I've got the notation clean that gives me insight into how to get on with the problem. And to maintain this concentration on the problem all day . . . of course you have to have meals and things and the children interrupt one too . . . but that is hard work. And like most things it's 90 per cent hard work, maybe 10 per cent inspiration and motivation. Just like any other job, I guess.

'But, if we just take the little problem that you were trying to solve, I think it's quite hard for the non-mathematician to understand why someone should spend six or seven years of his life working on such a very abstruse and very abstract problem.'

I agree, it is difficult to understand why. But then I found that little problem continually fascinating, and I knew in my bones that it was important and would hold the key. Of course when you start research you don't know what the problems are or why they're interesting. And that's one of the things you have to learn. Mathematics is a very vertical subject: you can't possibly understand the next stage until you've mastered the previous one. And that's why school-children, when they do maths, have no idea what maths at university is; and the undergraduate has no idea what a research student is doing; and a research student often has no idea what the top echelons of the mathematical world are doing. Well no, that's not quite true. I think research students do get a glimmering, because they often contribute to the top strata of maths.

'That's very different I think, from other sciences. Do you think maths *is* different from other sciences?'

Oh . . . yes. Mathematics is fundamentally different from science in many, many ways. The obvious one that people usually point out is that in maths we try to prove things and in science you try to disprove things. In maths you try and prove theorems; in science you try to disprove hypotheses. But it's deeper than that. What dictates research in science is the world; you're trying to understand what's there. But what really dictates research in mathematics is elegance. Mathematics is continually refining itself. You get several problems and maybe several theorems are proved, and then some genius sees a connection between these and he or she proves a supertheorem, of which all these previous theorems are little bits. So immediately we chuck all the old theorems away into the rubbish basket and only remember the supertheorem. And therefore the whole of maths has become a lot simpler because of this. So mathematics is constantly refining itself, self-simplifying, whereas science isn't. Science, by collecting new experimental facts, has to be always complicating itself—then it has to rely on something like mathematics to simplify itself again.

'I don't think that's strictly true. I think that for example a lot of science *is* simplifying. I think when you get new unified theories, in, for example, genetics, it does bring coherence to an enormous number of facts and you don't have to bother about all those facts any more.'

Right. Genetics, in biology, of course fits that. But when you look at the unifying theories in physics, such as Maxwell's equations unifying electricity and magnetism, you can't possibly describe those without mathematics. So mathematics is the tool that you have to use for the unifying theories. But there's another difference I wanted to mention between mathematicians and scientists. The scientist has to take 95 per cent of his subject on trust. He has to because he can't possibly do all the experiments, therefore he has to take on trust the experiments all his colleagues and predecessors have done. Whereas a mathematician doesn't have to take *anything* on trust. Any theorem that's proved, he doesn't believe it, really, until he goes through the proof himself, and therefore he knows his whole subject from scratch. He's absolutely 100 per cent certain of it. And that gives him an extraordinary conviction of certainty, and an arrogance that scientists don't have. And sometimes when you get mathematicians coming into science they have a most unwelcome arrogance.

'You talk about mathematicians coming into science as if you really do see mathematics as quite different from science. Would you say the actual mental processes of a mathematician are quite different from those of a scientist?'

No. Often they're the same, very similar.

'It's just the subject matter that's different?'

Right. Just as I often see that the way my wife, Rosemary, makes jewellery is very much the same as the way I make mathematics. She's not a scientist, but she struggles to solve a problem, concentrates on it, then suddenly gets an insight into how to do something. Sees how to put things together in a way that will give a harmonious balance as well as technical feasibility. It's exactly the same as some of the techniques I use to prove theorems. And she dreams about it, and gets cross with it, in the same way that I dream about my work and get cross with it. And I'm sure that most scientists dream about their work and get cross with it and get insights. So all these kinds of human experiences of someone towards their work are very common, I think.

'But you don't really think about experiments; or do you?'

Trying out things on the blackboard is like doing experiments. Most mathematicians don't do experiments in the real world, but the mode of thought is very experimental. You try all sorts of little tricks. You try and prove a thing one way and then try and prove the opposite. You're constantly experimenting one way or the other, or trying to invent a counter example.

'But one thing that does seem to be different is that you're doing it alone. You never say "one discusses it with a colleague".'

Ah well, I was talking about discovering or proving things by yourself. There is also an enormous amount of mathematics which is to do with other people. After you have proved something you have to explain it to people, and conversely you are constantly learning from others. And that's absolutely vital if you're going to keep up with the rest of the world, or keep fresh. There's a myth that mathematicians die out after the age of 35, but my best work has been done after the age of 35. That is because I work in a gregarious environment here at Warwick. The myth has arisen because mathematics is the devil's own job to read. It can take a day to read a page if it's an awkward page. And so to read a long paper can take you a month. It's terrible! But if you meet the author, and you get him to a blackboard and say, 'For Pete's sake, tell me what this is about', then he can graft it on to your knowledge. You just happen to know maths from a particular angle and he can go back until he finds your little angle and then with a few sketches and intuitive remarks he can probably communicate that month-long paper in five minutes. It's absolutely astonishing the tiny proportion of time it takes in talking to get an idea across, compared with reading. And no other subject has this in so extreme a form. So that's why it's absolutely vital to get together constantly with other people in your field.

'Can mathematicians in quite different fields talk to one another? Can you talk to any mathematician in any field?'
No, no, absolutely not. Most mathematics is a closed book to me.

'Did you say that *most mathematics was a closed book to you?*'
Yes, right.

'I find that very surprising; I would expect any reasonable biologist to be able to explain his work to me. But you're saying that's just not true of mathematics?'
Certainly not, It's the technical details. I'm pretty confident that with most mathematics, if I gave myself time, I could certainly understand it. But to get into a subject, a whole subject, which I was not familiar with, would take, say, three years. There just isn't time for me to get to know most mathematics.

'So in those three main areas, geometry, algebra, and analysis, there are subdivisions which lead one off on to very different and, to outsiders, mysterious paths. And that comes back to the point about it being a rather private activity. I mean how many people are there working in a small field of mathematics . . . are they talking to hundreds of other mathematicians, or tens, or fives?'

Always hundreds. Well ten or five in technical detail, but hundreds will know the broad pattern of what they're doing.

'And will care?'

Yes, well, if only five care then you're really out on a limb. You may be a great pioneer of course, or you may be doing something rather stupid.

'Is the opinion of colleagues important? Is it a competitive field?'

It's highly competitive. I mean, once somebody's got a proof, that's it. You may find a different or a better proof, but the theorem is always going to be known as *their* theorem. Although it's not like theoretical physics. Theoretical physicists feel that everybody has to work on the latest fashion, whereas there are so many different parts of mathematics you can work quietly away by yourself for ten years. And you may produce a theorem that will then turn the whole steamroller towards you. But, of course, you won't know whether it will until it happens.

'So it's really a question of trying to prove things. But what I think is difficult for the outsider is that you're not in touch with the real world. It still seems to me a sort of private game.'

Yes, yes it is in a way. I agree that it's a private game. But it has a seriousness that makes it more than a game. Otherwise, you might just say 'what's the difference between maths and chess?'

'Or poetry?'

Well, no, poetry tries to describe the world, tries to describe human emotions. It's true that the world stimulates mathematics. Mathematics can often go off on a tangent and then it's the world that brings it back on to the correct path again. There's a nice analogy—a nice story that René Thom tells—that when a baby is first born it babbles in all the phonemes of all the languages in the world, but after listening to its mother for a while it only babbles in the phonemes of her language. So we mathematicians babble in all the phonemes of possible mathematics and we should listen to Mother Nature to find out what mathematics is really relevant in this world.

'But do pure mathematicians really think about relevance?'

I think the best ones do. The best ones have an overall view of the subject and an overall view of which parts are 'mainstream'. We often talk about mainstream mathematics and refer to little sidebranches. Now what the heck is mainstream mathematics? It's whether it's got enough connections with all branches of mathematics and with those most important branches of physics.

'I'd always got the feeling that pure mathematicians didn't really care very much about applied.'

Well it's true that most of my pure colleagues don't care two hoots about applications. In fact they're rather scared of them, they steer clear of them.

'So why do they do mathematics . . . what's the pleasure that they have from the mathematics?'

Well, they do it because of its intrinsic beauty. I mean, it's constantly suggesting new problems. Once you get into a subject it's a very coherent structure and you're fascinated by how little the axioms are, on which you can build such a marvellous coherent structure.

'It sounds terribly self-indulgent, mathematics. If you say it's really just looking for beauty and elegance that may be lovely for the mathematician, but it's very much a self-gratifying activity—you're almost doing it purely to please yourself. Is that very unfair?'

No that's quite true. Perhaps we shouldn't say that they only care about elegance. They care about the subject. After all, I think it's a very noble subject. It's probably one of the oldest and noblest of man's activities. And I would identify, historically, with that very long tradition.

'You said noble—that's a strange word to use.'

Yes, noble. I've tried one or two other jobs and I've been amazed at the lesser feeling that I have about them. I don't know, you have certain things inside you which you feel, deep down, are important—your religious values, perhaps. Well, maybe in my assessment of the importance of things in the world, both the teaching and the mathematics itself come into that category. And I think you'd feel that about science too, wouldn't you?

'Yes I do'

Yes, well I feel that very strongly.

'Why do you feel that? Why do you think it is so important—not just for historical reasons?'

My real justification is that mathematics is a natural and a fundamental language. It may well be that it's a property of human beings, that only human beings can think maths. But I think it's probably true that any intelligence in the universe would have this language as well. So may be it's even greater than . . . no, not greater than, but more universal than, the human race.

MOLECULES
OF LIFE

DOROTHY HODGKIN *was born in 1910 and is*
Emeritus Professor at the
University of Oxford.

CHAPTER 6

FINDING WHAT'S THERE

─◦◉◦─

DOROTHY HODGKIN
Chemist

IN 1964 the *Daily Mail* carried a headline, 'Nobel Prize for British Wife'. The prize was for chemistry, and had been awarded to Dorothy Crowfoot Hodgkin for research on the structure of biologically important molecules including penicillin and vitamin B12. Using the technique of X-ray crystallography she had coaxed from these molecules the minute details of their three-dimensional structure, to the extent that the exact position of each atom was known. In 1969, she went on to solve the structure of insulin.

Professor Dorothy Hodgkin, Nobel Laureate, Fellow of the Royal Society, Emeritus Professor at the University of Oxford and the first woman since Florence Nightingale to have the Order of Merit conferred upon her, was born in 1910, in Cairo, where her father, Dr J. W. Crowfoot, was in the Egyptian Ministry of Education. Both her own family and that of her husband, Thomas Hodgkin, had a long tradition of intellectual achievement and social responsibility, and she has combined her distinguished scientific career with an active commitment to the cause of world peace. She has now officially retired and I went to talk to her at her home in a small rural village about 30 miles from Oxford. She looks just like the famous portrait by Brian Organ which hangs in the Royal Society, and her hands, crippled by arthritis since she was a child, are familiar too, from the drawings by Henry Moore. We sat by the fire in the cluttered sitting room, with its faded rugs on the floor and books and pictures everywhere. She is still very active, and divides her time between writing up research papers and Pugwash, an international movement of scientists working for peace. The delight she takes in her chosen field is infectious.

X-ray crystallography is a way of studying the structure of molecules by shining X-rays through them. The beam is scattered, or diffracted, by the atoms in the molecule and registers on a photographic plate as a pattern of spots of varying intensities. That this pattern of spots could be used to determine how the atoms were arranged in the molecule was the brilliant insight of a 22 year old Cambridge student, Lawrence Bragg, who three years later, in 1915, became the youngest person ever to receive a Nobel prize. The technique brought about a revolution in physics and chemistry, and also, more dramatically, in biology. Not only did it make Dorothy Hodgkin's achievement possible, but it also led directly to the solution of the structure of the genetic material, DNA.

Bragg worked with relatively simple, inorganic crystals such as salt. When the technique was applied to biological molecules, which are larger and more complicated, the diffraction patterns were, not surprisingly, also more complicated. Their interpretation became a highly skilled and immensely time-consuming occupation involving painstaking measurement, complex and lengthy calculations, and no small measure of intuition. When Dorothy Hodgkin began her research career, the ground rules for interpreting the X-ray data had still to be worked out.

Among the leaders in the field was Desmond Bernal, later to become Professor of Physics in the University of London. He was pioneering the application of X-ray crystallography to proteins, the most diverse and important chemical components of cells, at the Cavendish Laboratory in Cambridge. In 1932, after graduating from Somerville College, Oxford, Dorothy Hodgkin went to work under him. Together they obtained the first successful X-ray photograph of a protein single crystal. This was a major achievement, for not only does the skill lie in taking the actual photographs, but also in growing the crystals in the first place. It's not a matter of following rules, but of almost alchemical skill. Dorothy Hodgkin's future achievements were to depend both on her talent for growing suitable crystals, and on the intuitive and dogged brilliance she brought to the study of the impenetrable spots.

Two years later, she returned to Oxford. Here, just before the Second World War, Howard Florey and Ernst Chain began trying to isolate the actual antibacterial agent from the mould studied by Fleming. This fortunate set of circumstances gave her the opportunity to start work on penicillin as soon as sufficiently pure samples were available. It was a major undertaking, but by 1945 the structure was solved. Soon afterwards, in 1948, vitamin B12, the factor which prevents pernicious anaemia, was isolated, almost simultaneously, by two British and American groups. This was a more complex molecule than penicillin, and even with the aid of one of the first electronic computers, its structure was to occupy her for the next six years. Then came insulin, a still more complicated compound, which had preoccupied her since the beginning of her career. Its chemical structure—the number and order of the chemical units from which it is composed—had been worked out by Fred Sanger, in Cambridge, in the early 1950s, and won him the Nobel prize for chemistry in 1958. But the monumental task of determining the exact configuration of the constituent atoms still lay ahead. Dorothy Hodgkin took it on.

———

I'm really an experimentalist. I used to say, I think with my hands. I just like manipulation. I began to like it as a child and it's continued to be a pleasure. I don't do very much experimental work now, but I get something of the same pleasure from going through the maps indicating the position of the atoms that result from the calculations that are carried out.

'I hadn't thought of crystallography as being an experimental subject.'

Well, it does involve experiments, usually, because you often have to modify the crystal in order to get understandable results from the intensities of the reflection of the X-rays.

'Now, to a non-X-ray crystallographer the reflections that you get from the X-rays look a little ordered, but it's very hard to see any structure in them. Is it just a logical process to interpret them or is there a great deal of intuitive skill?'

Of course now you can just interpret them by putting the photographs through a machine, and letting the machine place the reflections and measure their intensities and pass them into a tape full of numbers which you can put into your computer. It wasn't like that when I was young and it isn't what I think about. I'd start off any crystal structure operation by taking the photograph myself and looking at it and seeing straight away what there is about the structure that I can tell immediately from the distribution of the reflections on the photograph. I admit that I don't like some modern improvements which cut out photographs almost altogether and put everything through a counter. I got a lot of pleasure myself out of just looking at the photographs and guessing the answers even if one guessed imperfectly and wrong. Also some photographs are really very beautiful you know.

'So you had a great skill in being able to go from those two-dimensional, ghost-like pictures to a three-dimensional object. Why were you so successful?'

Well I don't know that it requires all that skill if you know the lines on which these things work. It was a great advantage to start early. I mean one gets a certain amount of notoriety from being the first person to do things which anybody else really could have done. What I find difficult to know is why more people didn't take up this particular method of attacking problems at the same stage as we did. It seems to me that once W. L. Bragg had taken the first step, the chemists and physicists should have realized much more than they did that this was a tremendous opportunity. But for those who came in at the early stages there was so much gold lying about that we couldn't help finding some of it.

'Your pleasure, you said, comes partly from handling crystals. Is this something that developed very early?'

When I was quite young, I think I was ten at the time, I went to a small PNEU class in Beccles. PNEU stands for Parents National Educational Union, and it was founded by a Miss Mason of Ambleside to improve the education given by governesses, in a private way, all over the country. They produced small books that would enable the governesses to introduce their pupils to the different sciences in turn. So the small book on chemistry began with growing crystals, which I think is quite a common way to begin chemistry, growing crystals of copper sulphate and alum. I found this fascinating and repeated the experiments at home, when we had a home, which was the following year. My father and mother had been abroad most of the war and came home to look for a house for us to live in so that we could settle down near the local secondary school for our further education.

'So you made a little laboratory at home?'

Yes. I did go on crystal growing, and then, when I knew about the elements of analytical chemistry, I also used to carry out analyses on a collection of minerals. Now how I came by this is quite a nice story which perhaps illustrates the situation. My father and mother, as I said, worked abroad in the Sudan, and when I was 13 they were just about to retire. They thought it would be interesting for us children to see how they lived out there and so they took the two eldest of us away from school for a term to stay in Khartoum with them. We didn't do very much in the way of lessons but my mother took us about with her to see the different things she was interested in. One of the visits that we paid was to the Wellcome Laboratories and we first of all went to the medical one. Then, next door, was geology. The geologists there had just brought some little tiny pellets of gold back, and to amuse us children, they showed us how they got these by panning the sand from the bottom of streams. Of course this started me off thinking why shouldn't we find gold. So we went and panned the sand at the bottom of the little water channel running through our garden, and found a black shiny mineral. Now, I had already made friends with the chemical section of the Wellcome Laboratories. Its head was a particular friend of my father, Dr A. F. Joseph, whom we called Uncle Joseph. I went across to him and said 'Please can I analyse this mineral and find out what it is?' I guessed and told him I thought it might be manganese dioxide because it was black and shiny like manganese dioxide. So he helped me try the tests and, of course, it wasn't manganese dioxide. It was ilmenite, which is a mixed ore of iron and titanium. After that, he gave me the proper sort of surveyor's box with little bottles of reagents. One could carry it about the country and test for different elements in the minerals one found. It had a sample set of minerals in little tubes, so when I got home I used quite often to try out experiments and see whether I found the things in these little tubes that were supposed to be in them, according to the books. Then on his advice I bought a very large and serious text book of analytical chemistry, and continued this interest all through the years that I was at school.

'So you really were a crystallographer at heart from the beginning?'

Well, again, there's a quite interesting thing about that period. You see, my mother was very much interested in my choice of subject. She approved of it. She and my father were not scientists at all, they were both archaeologists as far as they had a profession, though they were at that time working mainly in education. She bought me, amongst other things W. H. Bragg's books based on the subjects of his Christmas

Lectures to children at the Royal Institution. If you come across them, they're very good and still perfectly readable. One is called *Concerning the Nature of Things* and the other one is called *Old Trades and New Knowledge*. My mother was particularly interested in *Old Trades and New Knowledge*. I think she got it because she was interested in weaving and potting and things of that kind. But the one *Concerning the Nature of Things* describes the X-ray diffraction of crystals and has in it the words: 'by this means you can "see" the atoms in the crystals'. And so I really decided then that this was what I would do. It was very exciting.

'Now when you started work you began with insulin, but you didn't solve it straight away.'

No, no, good heavens no. The first measurements on insulin, like Bernal's first measurements on pepsin which I was a little involved in a year before, were wholly ahead of their time as far as there being any conceivable chance that we could work out the structure. We were both really totally inexperienced in even simple structure analysis at the time. We were faced with an enormously complex problem and though, right at the very beginning, Bernal suggested the way in which it could be solved, it seemed to me that no way was I, at the age of 24, going to set out on that path without trying the proposed method on very much simpler problems first.

'So did you abandon insulin then?'

I never wholly abandoned it. I left it, yes, but in a curious way I didn't even really leave it. I went on doing the sort of things that would eventually have to be done, but in rather imperfect ways. Very slowly and gradually during the war, doing the measurements out in this house where I brought my little child to be safe away from possible bombing. I knew they weren't really good enough measurements to solve the structure, yet I couldn't help going on doing them somehow. But I put my real effort, when I was back in the lab again, into soluble problems. The first one was a carryover from the work that I was doing in Cambridge with Bernal, which was concerned with finding the structure of the sterols, and particularly of cholesterol. The next one was penicillin.

'Why did you choose penicillin?'

Penicillin was just historical accident. The work on penicillin began in Oxford just before the war and one of my friends at that time was Ernst Chain. In Oxford we go up and down South Parks Road, and going along South Parks Road one morning I met Ernst Chain in a state of great excitement having just been carrying out the experiment that is now famous. They had four mice which they injected with strepto-coccus and penicillin and four mice which they had injected with streptococcus alone. One group, the last group, died and the other group lived. And, as they were trying to isolate penicillin, Ernst was extremely excited, and said 'Some day we'll have crystals for you.'

'Now, when you chose to work on penicillin, was it that you really cared about penicillin, or was it that here was an important molecule which offered you the pleasure of finding out its structure?'

I think that both elements went into that particular operation. I mean nobody who lived through the first year or two of the trials of penicillin in Oxford could possibly not care about what it was. But also it's difficult not to enjoy just growing the crystals.

'Now, the structure of penicillin was soluble, unlike insulin which you were still holding in the background. Was that because it was simpler?'

Yes, there's all the difference in the world between working out the arrangement of atoms in space of a small molecule in which you've got 16 or 17 atoms in the assymetric unit and one in which you've got a thousand.

'How complicated, then, was vitamin B12?'

That's intermediate. That's of the order of 100. We didn't know it was intermediate when we started. Vitamin B12 came about through the fact that I got to know the people in the pharmaceutical industry rather well through the penicillin work, and Dr Lester Smith of Glaxo was working on the isolation of vitamin B12 just after the war. He got crystals first in 1948, within a week or two of crystals being obtained in America by the Merck Group. But to get its structure was again a long process because of the number of atoms in the molecule. The fact was, of course, at that moment we knew nothing, but nothing, about the structure, and what really held us up was just the state of computing in the world. You see electronic computers were being built, but when we started the beginning of the B12 work they hadn't been used at all in X-ray crystallography. They weren't really in a fit state. We did the end of the penicillin calculations on an old punched card machine and we brought this back again for the beginning of the B12 calculations. But it was very slow. A calculation which at the end of the story was taking, well, still a few hours on one of the early electronic computers, took us three months on punched cards.

'Did it matter that it was slow? Was it competitive? Were you frightened that somebody else was going to get the structure first?'

No, I wasn't really frightened that somebody else was going to get the structure. It was obviously going to be difficult and, actually, there was a rival fairly soon. Merck, the rival firm to Glaxo, decided that they needed a crystallographer to work on the structure and somebody had suggested a colleague of mine, John White, who had gone over to Princeton as a research fellow. I just took it as a compliment, since we had both been involved in the penicillin work. And although, of course, we were formally supposed to be rivals at first, after a time, when we were dreadfully bogged down, we went into collusion. I think we were regarded as wholly unreliable by the firms to which we were attached, and we ended up publishing jointly.

'At some stage you must have realized that you had talents or abilities that set you apart from your friends and colleagues. When was that?'

I don't think it was so very obvious, you know, because, in a curious way, of the sketchiness of my education. In the early period, before the age of 11, when my parents came home and I started secondary education, I had moved from one school, one little sort of private school, to another, and one year we'd spent actually being taught entirely by my mother, which was a very fascinating time. So when I first went to secondary school, I was rather behind, if anything. I was *terribly* behind in arithmetic, and it was only at the end of my time there, at the very end of the last year, that I was first in the form. One of the other girls was also very good indeed, and she was generally first. Actually, she did better in chemistry in school certificate than I did. She was very, very good.

'So what were your special qualities? Why didn't she go on and do all the things that you did?'

I've always thought that her case is really a case history for the problems of girls' education in this country. You see, a girl's future has depended a good deal on ambition and the advice that the young get given. She just didn't think in terms of going to a university, although there's no reason at all why she shouldn't have. I think in terms of the present organization, she would have done so, and would very likely have ended up in research.

'Are these same attitudes reflected in the headline in the *Daily Mail* which read: "British Wife gets Nobel Prize?"'

Oh, I thought it said 'Grandmother'!

'But do you mind that sort of thing?'

I didn't mind that one, no. The one I was slightly worried about concerned the penicillin work. The headline on my election to the Royal Society read 'Mother was first'. I wasn't quite sure that my chemical colleagues would have really appreciated that.

'Have you felt strongly about the position of women in science?'

No. I think it's because I didn't really notice it very much, that I was a woman amongst so many men. And the other thing is, of course, that I'm a little conscious that there were moments when it was to my advantage. My men colleagues at Oxford were very often particularly nice and helpful to me as a lone girl. And at the time just after the war, when there was an air of liberalism abroad and the first elections of women to the Royal Society were made, that probably got me in earlier than one might have as a man, just because one was a woman.

'You refer more to the Royal Society than you do to your Nobel Prize. Was it a surprise to you being elected to the Royal Society when you were?'

No, because I did know that it was not unlikely. When I first thought of it, I thought of it as something that I would much rather have than a Nobel Prize. You see, the thing that one missed out on as a woman was a certain amount of the absorption into scientific societies. In Oxford, the worst thing was that the Alembic club, which was the chemical club of the university, was not open to the membership of women when I was an undergraduate. We were allowed to go to the general meetings but not to the small meetings discussing research week by week. This I minded a great deal. I think I saw the Royal Society as one stage on from that, as a society of people who like talking to one another about scientific affairs, which was what I wished to do. And Nobel Prizes were a little bit out of my knowledge. As a young girl I didn't know a thing about Nobel Prizes. I wasn't particularly ambitious. I just liked working in this particular kind of field. I didn't imagine myself making enormous discoveries.

'In retirement you've got more involved with Pugwash. Have you always been involved in political activities?'

Yes. Generally socialist, and as you might say, peace loving. This I really inherited from my mother. My mother had four brothers who all took part in the First World War. Two of them were killed outright in France, the third died of the effects almost immediately after the war, and the fourth lasted a few years longer. And she became very much concerned with, particularly, the League of Nations. One of the early deeds she did for me was to take me to Geneva to one of their meetings. It was in 1924, I think, and it was an extremely interesting and important meeting. Nansen came and appealed for aid for the Armenians. I've always been glad I had that particular occasion as part of my history.

'Do you think, though, that scientists have a particular political responsibility as distinct from that of the ordinary citizen? Are you political as a scientist or as a citizen?'

Oh, more as a citizen I think. It's obvious that in certain aspects of one's scientific research one can't help feeling that one should try to do work that isn't going to harm anybody. But apart from that I think it's more important as a citizen than as a scientist.

'Are you optimistic at all?'

Well, I'm always optimistic when I meet people around the world, and particularly in an organization like the Pugwash Movement, because one has so many friends in spite of everything.

'Your other interest is archaeology. From your mother too?'

From my father and my mother, but very particularly from going on one archaeological expedition with my father and mother before I went up to college. I went with them to Jerash in trans-Jordan where they were excavating Byzantine churches and I very, very much enjoyed this time. It has, of course, a certain similarity to chemistry and crystallography; I mean you're finding what's there and you aren't controlling your situation. You're finding what's there and then trying to make sense of what you find.

Francis Crick *was born in 1916 and is J. W. Kieckhefer Distinguished Professor at The Salk Institute in California.*

CHAPTER 7

JUST GOSSIPING

—⊸•◉•⊶—

FRANCIS CRICK
Molecular biologist

'I HAVE never seen Francis Crick in a modest mood. Perhaps in other company he is that way, but I have never had reason so to judge him. It has nothing to do with his present fame. Already he is much talked about, usually with reverence, and some day he may be considered in the category of Rutherford or Bohr.'

That was written by Jim Watson in 1968. It opens his book *The Double Helix* which describes how he, a brash young American, came to Cambridge in 1951 and, with Crick, solved the structure of that most famous of biological molecules, DNA.

The discovery that the structure of DNA was a double helix depended largely on the technique of X-ray crystallography. Lawrence Bragg, who had been among the first to realize that shining X-rays through a crystal produces a pattern of reflections from which the arrangement of atoms within the crystal can be determined, had succeeded Lord Rutherford as Director of the Cavendish Laboratory in Cambridge in 1937. This was a surprise as everyone had expected the prestigious appointment would go to another nuclear physisicist. However, it was a prophetic decision. The Cavendish was to nurture the pioneering work of Max Perutz and John Kendrew on protein structure, and give the newly arrived Jim Watson a desk in Francis Crick's office. It was in this laboratory that the celebrated model of DNA was built, and the far-reaching possibilities of its structure revealed. It was not, however, in Cambridge that the crucial X-ray photographs were taken. That was in London, at the Biophysics Department of King's College, where Maurice Wilkins and Rosalind Franklin were already working on the structure of DNA. Rosalind Franklin died in 1958, but in 1962 Wilkins, Crick, and Watson shared the Nobel Prize in medicine and physiology.

DNA, deoxyribonucleic acid, is the molecule that carries genetic information from one generation to the next. It is an enormously long, and deceptively simple, molecule, made up of sequences of four basic units, or bases, strung together. It is packaged into chromosomes which sit in the nucleus, and rule the life of the cell. At the time when Crick and Watson were attempting to work out its structure there was some evidence that DNA was material from which genes were made, but it was widely thought that it could not, on its own, provide all the necessary information. There was also the question of how the genetic material duplicated itself. Every time a cell grows and divides, the daughter cells must get their full complement of genes. How was this accomplished? With solving of the structure of DNA came the key to both these problems, and genetics and the whole of biology was revolutionized.

The double helix at once suggested how the genetic material makes copies of itself. The DNA molecule is made of two strands that coil round each other—the double helix—and the bases in each strand pair with a different base in the sister strand according to a fixed rule. To replicate, the strands unzip and each is used as a template on which to assemble another strand from a pool of free bases within the cell. In this way each unzipped strand gives rise to a new and identical double helix. At the end of the short letter to *Nature,* the scientific journal in which they published their results, Crick and Watson wrote in an almost coy aside: 'It has not escaped our notice that the specific pairing we have postulated immediately suggests a possible copying mechanism for the genetic material'.

Then there was the other question. How did the DNA molecule alone carry all the genetic information necessary to determine everything from the colour of our eyes to the detailed chemical functioning of every cell in our body? Crick, together with another collaborator, Sidney Brenner, was one of a small, international group of scientists who began to try and answer that question.

We now know that genes exert their effects through their control over the synthesis of proteins, which are at the heart of the machinery of life. Enzymes, which facilitate all the chemical reactions in the cell, are proteins; all the main structures of the body are built on a protein framework. Proteins are essentially long strings of units called amino acids, and at the same time as Crick and Watson were thinking about the structure of DNA it was becoming apparent that the nature of a particular protein depended on the order of these amino acids and that for any given protein the sequence was always the same. Soon it was clear that the DNA molecule was not only capable of forming a replica of itself, but also of bringing about the formation of a specific protein. This presented the problem of the genetic code. Somehow, the four bases of the DNA molecule must specify the twenty amino acids found in proteins rather as morse code, with its different combinations of dots and dashes, specifies letters of the alphabet.

But as Crick and Brenner were to discover, a major stumbling block in solving the problem was the sorting out of the role played by another nucleic acid, RNA. Protein synthesis takes place in small particles in the cell called ribosomes. It was known that these ribosomes contained RNA, and it was thought that this RNA somehow carried the code for the protein. Only with the recognition that there was *another* RNA, messenger RNA, which carried the information from the DNA molecule to the ribosome, could the code be cracked.

Crick is now in his early seventies. A tall, ebullient Englishman, he has left Cambridge and works at the Salk Institute in California. His own interests, like those of many other molecular biologists, have shifted to the nervous system, and he recently published a theory of dreaming, in which he proposes dreaming as a process of *un*remembering, drawing an analogy with the necessity to clean out a computer's memory. As ever, his ideas are unconventional and characterized by their novelty and clarity.

He still returns to Cambridge every summer, spending a frantically busy few weeks meeting up with old friends and colleagues. I caught up with him very early one morning at his home, the Golden Helix, which sports a large double helix attached to the outside wall. We sat opposite each other in the dining room, which is just below street level, at a table covered with a red and white check cloth. The story of the discovery of the double helix has been told elsewhere—not least in Jim Watson's book. I wanted to go back further, to try and find out what had shaped Crick's life up to that point, and I also wondered how he had coped *afterwards*. How do you avoid the rest of your life becoming a kind of scientific anticlimax when you have made such a fundamental discovery?

———

As far back as I can remember I must have been interested in science, and

I particularly know this because my parents were not. But they bought me something called *The Children's Encyclopaedia* of which I still have the bound copies. Whether they came out in serial numbers or whether I had them bound, I don't remember, and I don't recall very much about what interested me. I recall looking at things in astronomy and so forth; but I was apparently interested enough in science to say to my mother, 'Isn't is a pity by the time I grow up everything will be discovered and there'll be nothing left to find out', which, as we know, of course, isn't the case at all. So that's, I think, how I began to be interested in science. It must have been very early.

'Now, I know you trained in physics at University College London. Why did you choose physics and not biology?'

Well I didn't really choose physics, in the sense that in those days you either did physics or chemistry or mathematics, and the chance of your doing biology wasn't very great unless you were particularly interested. I was just slightly better at physics than at chemistry or mathematics.

'Was it always with the aim of doing research?'

I suppose it was—although I didn't put it to myself in those terms. I thought I would get a job, and the job, I thought, would involve some sort of research.

'Were you a successful physicist?'

Not really. In a sense I had only just begun. My professor gave me a problem which was *totally* unsuitable for me and I hadn't got enough sense to realize it. I had to build a machine to make a precise measurement, and anybody who knows me knows that that is not the thing to arouse my enthusiasm. Actually I enjoyed it more than I expected because I was *doing* something, instead of just sitting there learning things. It was quite fun at University College, building this dreadful machine to measure the viscosity of water at temperatures above 100°C. You can see that it wasn't very exciting. But the war came just before I had begun to gather results. The laboratory was closed and, by good fortune, a landmine fell and destroyed the whole thing so I wasn't tempted to go back after the war and take it all up again.

'So what did happen after the war?'

I was in the Admiralty, of course, and during that period I designed magnetic and acoustic mines and had rather a good time in some ways because I could do largely what I wanted. But after the war I really didn't want to go on being a scientific civil servant—as I was—for the rest of my life. The problem was that I didn't know what I *did* want to do. By that time I was about thirty, you see, and most scientists have, by that age, a subject and a number of published paper, so it's difficult for them to change course. They really have to make a rather deliberate attempt. In my case all I knew was a little bit about magnetism and something about hydrodynamics. I knew that wouldn't get me very far, so I decided I could throw it all away and start again.

'You hadn't published any papers by then?'

No, not at all. Not one by that stage. You see, I had only done just the couple of years building this apparatus and even if I had published a paper it would have been a terribly dull one. So I decided that this was a marvellous opportunity. 'Here you are at the age of thirty, you can go into what you like.' The problem was what did I like? Finally I noticed the kind of things I was telling some of my young naval officer friends about. I used to talk about antibiotics, for example, and it occurred to me one day that I didn't know anything about antibiotics; I was just gossiping about them. So I decided that the gossip test is a good one, that what you're really interested in is what you gossip about. I looked at what I was gossiping to people about in science and it boiled down really to two areas. One was the borderline between the living and the non-living, and the other was the way the human brain worked. So I decided it was going to be one of those two, and after much thought and worrying about it—because, of course, I knew nothing about either of the subjects—I decided it had better be molecular biology. It was nearer to what I knew.

So I made this decision and then a very awkward thing happened about a week later. I was offered a job by Hamilton Hartridge, an eminent neurophysiologist who was setting up a unit to work on the eye. He asked me to come along and see him and offered me a job. This was a difficult decision. But I said to myself 'Look, this is the main decision you make in your life. Last week you decided it was molecular biology. You mustn't be deflected by small matters.' So I turned it down and that is how I decided to work on molecular biology.

'Molecular biology hardly existed as a subject at this stage. How did you set about it?'

I went and saw Massey, a physicist with whom I worked during the war, and he introduced me to A. V. Hill, the physiologist. Hill said 'You must go and see the Head of the Medical Research Council. And you ought to go to Cambridge. You'll find your own level there.' So I went to see Mellanby, and the Medical Research Council gave me a studentship, and my family helped me a little. And I went down to Cambridge, visited one or two labs, and was taken on at the Strangeways Tissue Culture Laboratory. I had previously suggested to the MRC that I should perhaps go and work with Bernal, the crystallographer, but Mellanby said 'No, we can't have that.' I think he felt that I must get among biologists and not be distracted by people who were merely doing X-ray diffraction. But after a year or so I decided that perhaps X-ray diffraction *was* the better thing and Max Perutz had been set up with a Medical Research Council Unit so I joined him. By that stage I'd taught myself a little bit of elementary biology, but I still had to learn X-ray diffraction because even if I'd had it in my physics course, I'd forgotten it all. So I had to teach myself that as well.

'Yet you were going to solve the structure of DNA within about two years. That seems an astonishingly short space of time.'

Yes it was a relatively short time. But anyway, I started working on the X-ray diffraction of proteins, and I also became a research student again because I had not got a Ph.D. Now some people in Cambridge do not have a Ph.D. It's sometimes said that it is beneath the dignity of a Fellow of Trinity to have a Ph.D. so they don't bother to take one. But in my case they said 'Well, you really don't need one in Cambridge, but if you ever go to America they'll expect you to have a doctorate.' So I became a Ph.D. student and Max Perutz was my supervisor. But it did take a long time actually to get it because the subject was in a very undeveloped state. It was a miracle really that *anybody* could get a Ph.D. studying the X-ray diffraction of proteins.

'Was it the arrival of Jim Watson that really stimulated your interest in DNA?'

No, my interest in biological things was, of course, there all the time I had been at the Strangeways doing tissue culture. I'd been reading round the subject and already decided that the chemistry of genetics was what I was really interested in. But, of course, I did not know any of the people I was reading about. I had *heard* of Max Delbrück and Luria and people like that, and I'd seen their papers, but they were just names to me. And although Max Perutz and John Kendrew had some interest in biology, they were mainly interested in the crystallography. So when I met Jim it was like meeting somebody from the outside world, who had seen all these people and talked to them. And what was surprising was that within the first two or three days we realized that our point of view, or rather the point of view of the problems we were trying to solve, was almost identical. The things that we thought important we agreed on. Of course, he did not know any X-ray diffraction, and I had only read about some of the things he had worked on.

'Was it just chance that brought you together?'

It was essentially a chance meeting because the fact that he came to Cambridge at all was a matter of chance, as he recounts in his book. But there was no chance about the fact that once he had come to work with John Kendrew we should meet. We were bound to meet.

'Did you complement each other?'

Yes, we did in some ways. What really happened was that each of us taught the other what we knew. There was no question of dividing it up and my doing the crystallography and his doing the biology. My interest in biology was quite as acute as his and he was quite prepared and interested to learn about crystallography. In fact, by the time we finished I think he knew parts of the theory of crystallography—he was not much of an experimentalist, really—better than Perutz and Kendrew.

'Were those fun days?'

Oh yes, enormously so, yes. And in fact that whole period was fun because one was not disturbed by anything. You see, nobody knew me and hardly anybody knew Jim and if they did know Jim they thought he was a bit eccentric. I had practically no letters. If I came in in the morning and saw a letter on my desk, which happened once a month, well, good heavens as they say, that was a red letter day! Somebody had written me a letter! Not this business of going in in the morning to a horrible pile of mail. We didn't travel very much. A few people came to see us in

Cambridge. We were not under a great pressure to produce papers. I did produce a few papers as part of my thesis which was just sort of going along, but you didn't have that feeling 'Oh! I must publish something or I won't get a job.' So you could just concentrate on the matter in hand. If it was a nice day you said 'Well, let's go out on the river' and we would go punting on the river, and probably go on chatting about the problem of genes, or whatever, while we punted. It was blissful in a way, and I have tried·very hard to re-create these conditions but all I can say is everybody is against me.

'Did you recognize from the beginning the importance of what you were doing?'

Well let's put it this way. I think we recognized the importance of finding out what the gene was and what it did and how it replicated. We were absolutely clear about the importance of that. You must realise it was not so obvious at the time that the gene was made of DNA. In fact there was no evidence at all, in spite of what people now say, that the gene was made solely of DNA. There was evidence that DNA was part of the gene. But it could have been DNA plus protein which was the gene. We hoped that learning about DNA would tell us something, but I don't think we had much inkling that such a simple structure would tell us so much. That was the lucky bit. We were looking for the answer to the problem and I think we were lucky in the way we stumbled on it. It is the only case where molecular structure has guided fifteen years of research for example, just in one blow.

'What did you actually feel when you discovered the structure?'

Well, you understand it came extremely suddenly. It is almost true to say we went in one morning with only a sort of faint idea that there might be something. Within a day we could see it was likely, and once we had built the first rough model—which took me a couple of days with Jim hovering nervously in the background—it looked very probable. But we weren't quite as convinced as people would think we were now. First of all, we had not seen the X-ray data at that stage. We had only glimpsed it. Secondly, model building is not a very reliable thing and so on. But we saw the implications jolly quickly. So we were in a state of euphoria about it, but it was tempered by the fact that we were not quite certain it was right. As Jim said in his book, every time he heard me telling people what a marvellous thing it was, he would have a queasy feeling that he was making a very big mistake. Fortunately, it turned out we did not make a big mistake.

'It was an amazing and explosive discovery but it must have presented you with a very severe problem: what could you do after that?'

Yes, well, one had two sorts of choices really. One was to go on, of course, and establish the DNA structure properly, which, incidentally, was only really done fairly recently. We decided we wouldn't do that because Maurice Wilkins was very well placed to do so. Instead we decided we would go into the other aspects of the problem. We didn't choose DNA replication. We *did* choose to look at how genes expressed themselves, what we nowadays call the genetic code. We were reasonably sure it involved protein synthesis. So both Jim and I actually did experiments in the following years, though with mixed success shall we say. But we also did an awful lot of propaganda, getting people to see the problem, and talking to other people who were doing the experiments. And I think it was the right choice, looking back, not just to have stayed with the crystallographic aspects but going into the wider ramifications of the problem.

'Were you conscious at that time of your tremendous success. Were people looking to you for something equivalent? Was it a sort of albatross round your neck?'

No, no, not at that stage. I don't say it didn't become so later. I know it may seem absolutely incredible, but it was not until two or three years afterwards, when the American protein chemist Frank Putnam said something to me about it, that I realized it was the sort of thing they give prizes for. It had never occurred to me.'

'You mean you were never conscious that you might get the Nobel Prize?'

Not for several years. It never occurred to me, and Jim never mentioned it. He gives the impression in the book that he was thinking about it the whole time both before and after. But I have checked with John Kendrew and my other colleagues, and on no occasion whatsoever did Jim mention this to us. He must have kept it very quiet and to himself, and it certainly didn't occur to me. I had read C. P. Snow when I went back into research, and because of that I'd wondered 'Shall I ever be made a Fellow of the Royal Society, that splendid thing the FRS?' You remember, there was a character in one of his novels, *The Search,* who is never going to be made an FRS. Well I had decided that you shouldn't worry about that sort of thing, and if towards the end of my career I was made an FRS, that would be fine. Otherwise let's do it for the fun of it, you see. And this business of a prize, well, of course, once somebody mentioned it to me, I suddenly realized 'good heavens, it *is* just the sort of thing they give prizes for!' But even then I can't have been very conscious of it because, as you know, the announcement is made in November each year and they would always announce the prize before I got round to wondering who it was going to be that year. Eventually of course, it did come. I remember the occasion as one does, rather vividly. You're always supposed to wait for the telegram, you know, and not worry because of the journalists hovering about from Sweden . . . . There is a story that we filled a bath with champagne but that is totally untrue. It is traditional to serve champagne, so we filled the bath with *ice* and we put the champagne bottles in the ice . . . I think I did realize it was highly unlikely one would make another discovery which would have the same impact but, as one says, well so what? Let's go on and do interesting things and not bother about it.

'But, Francis, you were now very famous.'

Alas, I am now, but let me tell you, the year after the discovery of DNA I was not specially famous. I was living in Brooklyn, in fact. I think by 1959, when I gave some lectures at The Rockefeller and was a visiting Professor at Harvard, I began to realize that people knew about me because I was getting so many invitations to go everywhere. But I didn't think of myself as famous. I thought of myself as well known, and there is a distinction there. I was well known among *scientists.* The lay person had never heard of me, because it had not got into the textbooks. I can't remember the first time people began to show shock when they were introduced to me. That was considerably later.

'Did you publish much?'

I published a review called *On Protein Synthesis* in 1958, which I think was an influential review. What I intended to do, and never did was, of course, to write a book, but I was always too busy. I'm not really much of a writer, but I did want to write a book in 1954–55, saying what the implications of the structure of DNA were, and to discuss the idea of genetic code. But as it was, it was done by word of mouth and that curious organization the RNA Tie Club, founded by Gamow, and this one review article.

'The RNA Tie Club?'

Oh well, Gamow, you know, was the cosmologist. He did a number of other things but he was particularly good on the theory of the Big Bang. He also became interested in DNA and published a paper proposing that cavities in DNA could act as a way of assembling amino acids for proteins. He was a very amusing man. His idea of a nice afternoon was to sit on the beach and show card tricks to a pretty girl, roughly speaking, while he was drinking something. And he founded a very select club because it had to be limited to twenty members, one for each of the twenty amino acids. I think I was tyrosine and you were supposed to have a tie pin. There was a tie too, which we still have, made by a haberdasher in Los Angeles. And there were supposed to be four honorary members for each of the four bases and people wrote little scientific Notes. My most influential *un*published paper was for the RNA Tie Club. It is the one which essentially suggests there ought to be something like transfer RNA. I don't think we ever met as a club, but there was notepaper, and it did serve to improve the morale of the small number of people who were thinking about this way-out and, to some people's point of view, imaginary problem about what the genetic code was.

'You mean people did not really accept that there was a genetic code?'

Well, it's not that we had a lot of opposition, but I think there was a feeling that we were oversimplifying things which, of course, in a certain sense we were. But we were not oversimplifying the outlines. It's the details that nature embroiders. The outlines are pretty bold and those were the ones which we had more or less right. But we made one terrible, terrible bloomer. In modern terms we would express it by saying we thought that the ribosomal RNA was the messenger RNA, and that held us up, oh, for several years. The penny dropped one day, one Good Friday I think it was. We went in to work that morning and we came out realizing that the ribosome was a reading head and that there was something called messenger RNA, which had not been discovered then, or rather it had been discovered, but people had not realized what it was. That was the second dramatic occasion that I remember, when, in a very short period of time, the whole of a subject began to look quite different. You see that single blunder held us up terribly. In 1958 I wrote a review about the genetic code. It was just so pessimistic because we could not make anything fit together. We had just this one, wrong idea. Once we got rid of that, I wouldn't say it was plain sailing, but everybody could get to work again, and the code came out within five or six years. Much more quickly than people expected.

'Now you are not in the mould, and certainly Watson doesn't paint you as such in his book, of the traditional Cambridge empiricist.'

Well, I don't know exactly what the traditional Cambridge empirist is. I am not especially a Cambridge person, although I have lived here for many years and been associated with the University, and even for a period been a Fellow of a college. But as for being the empiricist . . . .

'Do you like doing experiments? Did you do many?'

I did, you know. I wasn't very good at biochemical experiments. I think I might have become so, but I had a little bad luck at the beginning and that put me off. One experiment I ought to have done I gave to somebody else and they ran the vital spot off the paper. Just one of those things. But when I was doing phage genetics, genetics of viruses that infect bacteria, I did do a series of rather beautiful experiments which proved by genetic methods that the DNA was a triplet code.

'So you don't think of yourself just as a theoretician?'

No. Not at all. I enjoyed doing those experiments and Odile, my wife, said I was never better tempered than when I was doing that. You know, doing theoretical work, especially this sort of theoretical work, is very demanding. You read away, and you think away, and you worry and so on, but you don't have much to show for it. It's not a discipline in which you can sit down in the morning and get results of some sort. It is not a case of doing calculations, and working a computer, and there's the answer. I don't work that way. I have to have a problem which nags at me for a long time.

'How do you do your theoretical work?'

Well, what *I* do is, not choose a problem, but to choose a subject, and then try and move around in the subject until I find an idea that yields; something that clicks together. I don't say I am going to try and solve such and such a problem, because it may turn out, especially in biology, that it's insoluble. The problem of how proteins fold up, for example. We showed great discrimination in not choosing that problem. It has not been solved to this day. We put that on one side. The way I work is to take a given area and to look at problems which look as if they might be tractable, as we thought the genetic code would be. So that's how one does it. And how the ideas come, well that's very unclear, except that you have to do a long period of reading, talking, and thinking, and you have the business of trying to assess people. One of the problems the theorist has is that his raw material is other people's experiments. The people are as important as the experiments because if you know the people you know the sort of mistakes they make, for example. That's much more easily found out by actually meeting them, even for a short time, than merely by reading the papers. So if I'm reading papers by Japanese scientists, for example, it is very difficult for me to form an opinion of what the people are like. But once you've met them and talked for a little while, you soon discover whether they are deep thinkers or very careful experimentalists just from what they chat about. So it is important for a theorist not only to interact with experimentalists, but also to meet those he doesn't work with so closely. You have to go to meetings and go round and see people for that reason. I don't say it's not important for experimentalists as well, but it is doubly important for theorists.

'You have had famous collaborations. Jim Watson, Sydney Brenner. What has been the basis of that?'

Friendship. Certainly you have to be personal friends. You have to have common interests and of course you have to be candid. This is perhaps the most important thing. You have to be candid without being rude, so that you can say something which sounds rather aggressive but the other person knows that that's just the way that you usually disagree with him. So if you say, 'Oh, that's all nonsense' he doesn't turn a hair. I have to be rather careful now saying to people 'it's all nonsense' because they take it like a hammer blow, and often I only mean it in the most friendly way. So with people who are not close collaborators I do watch what I say. At least, I try to. With collaborators you don't want to bother about that. You've got far too much to think about and you must, of course, try and attack the other person's ideas. Because it's getting rid of false ideas which is the most important thing in developing the good ones, and that's what the collaboration is most useful for. Each helps the other person to get rid of his false ideas and they do it by this dialogue. Also it's very important not to have the meetings too spasmodically. Not just once a month. You must be able to do something, stew over it for a day or so and then go back to it, and then talk over it again. Then perhaps go back to it the next week, so it's constantly recurring. That's why you need somebody who is there quite a bit of the time.

'Does this really mean that with these people you spend a great deal of time just sitting in a room and talking?'

Yes, that is how it happens. Of course, Sydney Brenner is an experimentalist. We did share an office for whatever it was, twenty odd years, but he was often in the lab. But then he would come in and he would tell me what he had been doing, or I would raise something I had been reading and we would just chat away.

'One of the things that characterizes your work and the way you speak is its tremendous clarity. You never say something that I find difficult to understand. Do you do that self-consciously?'

No, it's not self-conscious. It's because I think I'm rather simple-minded, and have to explain things to myself. It's the way I talk to myself that I talk to others. Unless I can conceptualize things in a fairly straightforward way, I cannot think about them myself. Most people, I would say, don't attempt to do it. They don't make the effort. But if it's your job to think about something, there is really almost no other way of doing it.

'Do you have any heroes or people who influenced you?'

Well, I was influenced by certain people, yes. I was influenced by Bragg and Pauling. Especially by Bragg. I remember on one occasion Bragg was going to give a lecture on a crystallographic subject I knew very well, and so I had rehearsed in my mind how I would give the lecture. When Bragg came to give the lecture it was totally different from mine.

I would have given all the details and gone into all the ramifications. Not so Bragg. He made about three simple points. It was a quite technical thing, yet the audience understood what he was saying. They remembered it. And I learned that lesson. That's what you should do in giving a lecture. You should not, except in a very small seminar, get into all the technical details.

Both Pauling and Bragg also taught me something about how to approach a problem. That you should not, again, get bogged down with experimental details. You should make some sort of bold assumptions, and try them out and so on. Many research students cannot do that. They even may have a discovery under their noses and they don't see it because they think it's too simple. They have been trained so hard to believe that science is hard work, and that you have to read the literature, and that everything must be done carefully and you must go away and struggle. The fact that you can discover something in a *week* which is much more important than anything you've done in the previous year is not always apparent to them!

SYDNEY BRENNER
was born in 1927 and is Director of the
Molecular Genetics Unit of the
Medical Research Council in
Cambridge.

CHAPTER 8

NO ZOMBIE BIOLOGIST

———∽◦◯◦∽———

SIDNEY BRENNER

Molecular biologist

THE advent of molecular biology has been likened by the scientific historian, Freeland Judson, to the 'Eighth Day of Creation'. It has revolutionized our understanding of living things in terms of how genes behave at the molecular level, and given birth to the whole new field of genetic engineering. Sydney Brenner is not only one of the founders of the discipline, but was also, until recently, director of one of its outstanding institutions, the MRC Laboratory of Molecular Biology in Cambridge.

The preparations for his career were made in South Africa where he took a Master of Science degree and qualified in Medicine at the University of the Witwatersrand in Johannesburg. In 1952 he came to England with a scholarship to study at Oxford. The timing was perfect. Fred Sanger was about to demonstrate that proteins had a precise chemical structure and Crick and Watson were shortly to discover the structure of the genetic material, DNA. These two events defined the basic problem of what was to become the new field of molecular biology: to discover the connection between the structure of genes and the structure of proteins. Five years later Brenner had moved to Cambridge and was sharing a room with Francis Crick. It was an enormously productive collaboration which led, among other things, to the discovery of messenger RNA and fundamental insights into the nature of the genetic code.

Brenner is a brilliant experimentalist and it was largely due to his exceptional talents that the existence of messenger RNA was discovered in a few hectic weeks in 1960. The possibility that information was carried from the DNA in the nucleus of the cell to the ribosome, where proteins are actually synthesized, by a 'messenger' had dawned on Crick and Brenner at a meeting in Cambridge that April. Brenner and the French biologist François Jacob devised an experiment to test the idea, which they set up two months later while both were working in Max Delbrück's laboratory at Caltech in Pasadena, California. But things went very badly, and with only a few days left they were sitting gloomily on the beach with Brenner, usually renowned for his wit and conversation, unaccustomedly silent. According to Jacob, Brenner suddenly leapt up and shouted 'It's the magnesium' and they rushed back to the lab, repeated the experiment with the addition of more magnesium, an almost trivial component, and the existence of messenger RNA was established. After six more months of hard work back in Cambridge, the work was complete.

A similar combination of inspiration and experiment lay behind the work on the genetic code. Since Sanger had shown that the insulin molecule was made up of a fixed sequence of amino acids it had become clear that the particular characteristics of a protein were determined by its unique sequence of amino acids. Brenner and Crick proved that the four bases of the DNA molecule were able to specify the twenty amino acids found in proteins by means of a triplet code, a different combination of three bases corresponding to each amino acid. They were also right about the so-called 'sequence hypothesis', which predicted that specifying the number and order of the amino acids was all the information required for protein synthesis. The long, string-like molecules fold spontaneously into the three-dimensional structures which determine the properties of each protein.

It is hard now to realize how different this view of protein synthesis, couched in terms of information flow, is from the previously held view. Before Sanger's work it was widely thought that proteins had no definite structure and the problem was to discover how the amino acids were joined up, and where the energy came from to hold them together. As late as 1947 the great geneticist Müller still thought of genes as controlling the *energy* flow for protein synthesis. The 1950s saw, in the terms of the scientific historian, Thomas Kuhn, a major shift in the paradigm away from energy and metabolism towards information and structure.

In the early days Brenner worked on a favoured experimental system—viruses which infect bacteria. These viruses, called phage, are among the simplest organisms which can replicate and code for proteins. They provided excellent material for attempting to solve genetic problems at the molecular level. Later, Brenner, like other molecular biologists, such as the distinguished American geneticist Seymour

Benzer, abandoned bacteria and phage to work on higher organisms. With the genetic code worked out, they wished to apply the techniques of molecular biology to more complex biological functions such as development and behaviour.

Benzer started work on the behaviour of the fruit-fly, *Drosophila,* which has always been a favourite organism for genetic study. Brenner, by contrast, in an amazing *tour de force*, worked out, from scratch, the genetics of a simpler organism, a nematode worm, in order to study the development of the nervous system. It is now the experimental organism of choice in laboratories all over the world. Such a reductionist approach simply did not exist when he began to study medicine in the 1940s.

————

When I went to university, you couldn't do science—especially biological science. You either did medicine or agriculture. So I did medicine. It was the only way I could get a bursary to go to university. But I wasn't very interested in medicine, and switched to science at the first opportunity, particularly when I discovered that I'd be too young when I graduated to actually receive the medical degree. I would have graduated at 20 and you had to be 21. So I took a year off to do a science degree. Then I took another year to do an honours degree, and then I took a third year to do a Master of Science degree. That was 1947. I then had a chance to go to Oxford, and I went round asking people what they thought I should study. They said the subject we think you're interested in is called cell physiology, or maybe biochemistry, but all the jobs are for people in medical schools and it is very hard to get a job in a medical school unless you have a medical degree. So after all that I went back to do medicine. I was a very reluctant medical student. I didn't like medicine. I think I'm a *things* man rather than a people man. So the moment I got my degree I said, well, thank God I don't have to do anything like that again. I never practised. Medicine was just a way of getting into what we would call molecular biology now, but which I thought of as 'living chemistry' or 'biological chemistry'.

'What was it about biology that intrigued you?'

I don't know, but for example, I can remember extracting pigments from plants, and finding out that they were pH sensitive. I must have been about 11 or 12 at the time, so I had a long interest in biology, and especially the chemical parts of biology.

'So when did you really begin working in molecular biology?'

Well, of course, the subject didn't exist until long after this. I was doing something I called cell physiology. When I went to England in 1952, I asked people's advice as to where to go. They said that since I was interested in something half way between physical chemistry and biology, they recommended me to go and work with Professor Cyril Hinshelwood, who was then the Professor of Physical Chemistry at Oxford. He had written a book called *The Physical Chemistry of the Cell*, and because the Principal of my university in Johannesburg was an Oxford man he arranged it. I also applied to go to the biochemistry department at Cambridge but they never even replied to my letter. That's the last time I ever applied to that department. But I went to Oxford, and that was how I got into working with phage.

'So you were working on the foundations of molecular biology, and it was obviously just the right subject for you. You were part of a group who achieved an amazing amount in what was really a rather short time. Why do you think that was?'

I think it was just that we had a certain approach to asking questions. You see, the foundations of molecular biology were laid down as follows. Up until Fred Sanger's determination of the sequence of insulin most people believed that proteins were statistical polymers. I'll never forget a talk Sanger gave in Oxford, at which Sir Robert Robinson, the organic chemist, said he thought that this demonstration that proteins had a chemical structure was *astounding*. Now this gave us the 'sequence hypothesis'. To me, the most important thing about the early generalizations in molecular biology was not so much the structure of DNA as the sequence hypothesis. This really confronted the fundamental question that Francis Crick posed. This was, knowing, as we did, that proteins got their functions from their three-dimensional structure, if all the DNA specified was the linear order of amino acids, how was that structure determined? I can remember going to meetings where people said 'Well, there'll be genes for folding up proteins' and so on, and we had just this remarkably simple hypothesis, the sequence hypothesis, which said that all you had to do was to specify the amino acid sequence and the folding would look after itself, and the energy would look after itself, and everything would be all right. Nothing to worry about. Now this was, I think, the blinding insight into the whole solution of everything in biology and many people couldn't see it. They said, 'Oh well, you know it's too simple a hypothesis.' Biologists were very much preoccupied with complexity and they never believed in their heart of hearts that all the structures that they studied, you know, feathers, eyes, would ultimately be explained in terms of the form of molecules. There had to be more information. Chemical sequences are not enough.

Now, you ask why did we accomplish so much? Of course, one has to give a rather personal view. I think, first of all, in the early days of molecular biology, it was an evangelical movement. Most people were against us. Most of the biochemists didn't understand the nature of the problems that we thought were interesting and important. They had a completely different set of attitudes. But we didn't feel too upset by this. We all had, in fact, healthy disrespect for the establishment, and a group of us throughout the world got on with the work. Given the structure of DNA and the appreciation of what it could do, I think it was the fusion of the chemical and the genetic approach to the problems of replication and gene expression that really allowed the subject to take off. It's hard to explain because the different groups had completely different views about things. We, for example, were very much embedded in chemistry. We didn't have any of this nonsense about whether there was a new physics—although people like Francis Crick were physicists—we simply realized that biology was a branch, if you like, of very low energy physics, and in this little corner of the universe there would be a rather special chemistry which would have to be looked at and worked out. The job was to find it.

'What I'm trying to understand is why you people recognized that. You would have thought that biochemists, by their very name, would have adopted exactly the same approach. Was it changing what Kuhn would call the paradigm?'

I can remember meetings at which it was impossible to get across to people the idea that the most important thing in protein synthesis was how the order of the amino acids got established. They said, 'That's not the important problem. The important problem is, where does the energy come from to join the amino acids?' Well, we have written, on many occasions, that the sequence is the important thing, and never mind the energy, it'll look after itself. And really, this is what this part of molecular biology brought. It said that the flow of information can be studied at the chemical level. I don't think biochemists actually understood the importance of information at that level. It wasn't information theory, it was the flow of messages, and we tried to seek for explanation in terms of the molecules.

'Having done a tremendous amount in molecular biology, you suddenly started working on the development of a worm, the nematode. Why did you make that change?'

Well, I think this happened in about 1961 or 1962 when we felt that, in general outline, the classic problems of molecular biology were solved. There was still a tremendous amount of work to be done, no doubt, and people would be beavering away doing it, but, we asked ourselves, did *we* want to do it? So we wondered, could we build on this approach? Having got this view that we could take quite complicated things like how viruses are put together, or the biochemical pathways of bacteria, and understand them in terms of genes and their products, we wondered whether we could do this with anything more complicated.

Now, I chose the worm because it was finite. It has a limited number of cells, and from the old classical descriptions going back to Goldschmidt, its development looked very invariant, with a fixed final number of cells. This property of invariance suggested to me that you might have something really precise to account for. You would have, so to speak, a very precise genetic programme, using those words very loosely, and you could really get your teeth into it. And the problem was finite in the sense that you could come to an end of the description of the worm's development. As you know, you can make exhaustive cellular descriptions of these animals using modern techniques. You can cut serial sections and photograph them in the electron microscope. You can find out everything at that level about every cell, where it comes from and where it is. I think it is fair to say we have done just that.

'Do you think it was the right choice? It was quite brave because you really had to start at the beginning with the worm didn't you?'

Yes, we had to start right at the beginning. Everybody asked why we didn't work on the fruit-fly, *Drosophila,* because people had already been studying its genetics for 50 years. I took the, perhaps arrogant, view that we could catch up with the genetics pretty quickly. With the worm we don't have many of the advantages of *Drosophila* but we do have a much simpler organism. You see, Seymour Benzer used to say that *Drosophila* with 10^3 neurones is half way in complexity. It is the geometric mean between the bacterium *E. coli* which has one neurone because it is one cell, and man, who has 10^{10} neurones. We used to say that we went one better, because the worm, with three hundred neurones, is the geometrical average between *E. coli* and *Drosophila*. So it's orders of magnitude is less complicated than the fly and you could hope to try and get pretty well-defined answers. Of course, we've even found the worm to be too complicated and have had to work on subsystems. Nevertheless, I think it has been very instructive because we have been able to answer certain questions.

'You invested an enormous amount of work in the worm. Do you think you have caught up with *Drosophila?*'

Well, I think we have the capacity to do so. We're far ahead of *Drosophila* on certain things. At the cellular level we know much more. But I think on the more genetic things they have a little bit of an advantage over us. So, no, I wouldn't like to say we've caught up with them. We may have passed them on a different road, perhaps.

'We are talking about it as if it is a competition. Was molecular biology a particularly competitive subject in the early days, and is it now?'

Well, I think molecular biology now is viciously competitive and that's largely because it's enormous. You can judge a field by looking to see how many people simultaneously discover the same thing. I can show you cases where the hit rate is three or four. Three or four groups discover the same things, and they produce papers which are almost identical. They haven't copied from each other but you know they're all looking at the same piece of nature with the same techniques. And that's a sign of an overpopulated science. Of course a sign of an under-populated science is one where nothing's ever confirmed by anybody else and you don't know whether it's true or not. So I think it is very competitive. I feel sorry for young people in it. I was amazed when I discovered there were actually textbooks on molecular biology. I was sitting behind some students in the bus once and they were swotting for their exams and one said, 'Do you think we'll get the genetic code tomorrow?' I was horrified that there were people cramming the genetic code like I used to cram the path of the facial nerve in anatomy. They were trying with mnemonics to remember whether UGA was this or that amino acid. I came away very sobered by the idea that if you help make a subject you never have to learn it. So I got a bit depressed then when I discovered that molecular biology had become an academic subject—that there were questions in exam papers about it.

Also, the literature is now enormous and you can only follow it by living in a sort of cooperative, a reading cooperative. Or, as they now are, xeroxing cooperatives. Because hardly anybody reads any more—people only xerox things. I once asked a student who had a big xerox bill whether he'd tried neuroxing some papers. So he asked me what that was. I said 'It's a very easy and cheap process. You hold the page in front of your eyes and you let it go through there into the brain. It's much better than xeroxing.' You see, most people will xerox things on the grounds that if they have the paper they can read it at any time. So they don't have to read it now. It's on the same grounds as someone in Rome never goes to see anything there—it's only tourists who see things. Perhaps the only people reading in a field are the ones who don't know anything about it and aren't working in it. The ones who are in it are just xeroxing the papers. I think one should produce a copy of a journal with nothing it it and see how many times it's xeroxed. But that's by the by. I do think it is a very competitive field. The amount of information is enormous. People have reached a specialization that is unbelievable. I have people in this laboratory who can't understand each other. There was a time when everybody prided themselves on being able to follow at a deep technical level everything in molecular biology. But there are very few people left in the world who can do it. As you know, just as all the societies have split up into interest groups, so has the subject.

'What do you think the competition means to people—or to you? Is it the pleasure of being the first to discover something or the way it will affect your career? Why is it so competitive?'

Well, I think a lot of the competition is because there's a tradition of competition in the field. There's a kind of mythology of the subject from Jim Watson's book *The Double Helix,* and everybody wants to model themselves on those people. So, I think there is a tradition of this. I think also the population has grown enormously, and the amount of resources has flattened off. The number of jobs isn't enough for the number of really first rate people competing for them. I think it's got to the stage that priority is very important. I've often wondered 'Did we feel this?'. I think had Francis Crick not existed I'd have never written a paper in my life. It was only Francis who made me write papers, because once I solved a problem I lost interest in it. But Francis used to lock me in a room and say 'You've got to write it up.' So most of the things were written under duress, in fact. We just took ages to write things up. We never rushed things. Now many things are being written up even before the experiments are done in some cases!

'But if you didn't write papers how would people know that you've discovered something?'

You wrote them a letter in those days. You'd say 'Look we have this.' For example, we would send a pre-print of our paper to lots of people. We had a mailing list. People would receive this pre-print and they would have a seminar on it. Everybody would be made to read the paper. Somewhere I have all the correspondence from people when we sent out our first paper on the acridine mutants, which was the work on the genetic code. Now it's transparently clear, but it was so hard to convince people of those facts. They thought it had to be wrong somewhere. We had people writing saying they've had three seminars on this already. I mean people communicated and offered comment and criticism.

'So you don't like writing?'

I just don't. I have writer's freeze. I used to write quite easily but now I'm tired of it. I think it's just the business of scientific writing. What you're compelled to do is very bad for you, because you become as concise and as technical and precise as possible. This means that you start to say 'Well, do I really need all those words?' And so one writes in a very dense, almost distilled style, in which all the edges and all the extra things have been chipped off. And it's very hard to get it like that. And I've now found I can't write freely any more, although I talk pretty freely. If I have it recorded and then transcribed, it's complete rubbish. No doubt you know that.

'Now you said that when you began to work on the worm you thought that the real problems of molecular biology had largely been cleaned up. In retrospect, considering the explosion in activity, do you think that's true?'

I think that molecular biology is just with us all the time, and things have been worked out in enormous detail of course. When I said that we understood protein synthesis in outline, we did, although an enormous amount of work has gone on since. I suppose that is a line of work we could have done. In fact, Francis Crick and I discussed whether our lab should work on the ribosome—the site of protein synthesis—and work out its structure, do the function of all its components, all the protein sequences, and so on. That's a programme several large laboratories have since carried out, although, in fact, we still don't know the structure of the ribosome and still cannot give a really accurate molecular description of how it works. It's still an unsolved problem. But we knew in outline that you needed ribosomes and something would move through them, and the rest just seemed details, and to have lost the challenge.

'What do you thing have been your special skills in solving problems?'

I think two things. For 20 years I shared an office with Francis Crick and we had a rule that you could say anything that came into your head. Now most of these conversations were just complete nonsense. But every now and then a half-formed idea could be taken up by the other one and really refined. I think a lot of the good things that we produced came from those completely mad sessions. But at one stage or another we have convinced each other of theories which have never seen the light of day . . . I mean completely crazy things. They're all buried in drawers now! So that was one thing. I also think I'm a pretty good experimentalist, at doing genetics. And so it was a combination of having these ideas, and thoughts on how you might implement them, and then following them up with experimentation, which I think has been my contribution.

'Do you miss Francis now?'

Oh yes, yes. I do see him from year to year, but that was a great 20 years in which all these things were discussed and worked out. I mean nowadays things are different. I saw Jim Watson the other day and he says you don't have to *think* in biology any more, you know you just have to go and *do*. And in fact I once thought of founding this new journal, called *The Journal of Zombie Biology*. It's not about the biology of zombies—that's a very interesting subject—it's for zombie biologists. Because that's all you have to do. You just have to wind yourself up in the morning, and go to the lab and just do things. And of course it's marvellous because many of the answers come from just doing things. Biology isn't a subject in which you can have great thoughts in the bath. Leo Szilard, the nuclear physicist, used to say to me that when he left physics and went into biology he could never have a comfortable bath, because no sooner did he get into it to think, than he had to get out to look up another fact. When he was a physicist he could lie in the bath and think for hours, but in biology he was always having to get up to look up another fact. The subject has changed, but I still feel that there are nice little corners of it into which one can go. I think what I would like to do is to have another three post-doctoral periods of research in different parts of the subject.

GUNTHER STENT *was born in 1924*
and is Professor of Molecular Biology in the
University of California at Berkeley.
He is sitting in front of a portrait of Max Delbrück.

CHAPTER 9

TELLING NATURE

—◁◦⊙◦▷—

GUNTHER STENT
Molecular biologist

GUNTHER STENT has a reputation for holding somewhat heretical and certainly provocative views about science. A great friend of his told me 'Gunther is almost always wrong, but he is always interesting'. Although a leading molecular biologist he is unusual in being, in a sense, anti-reductionist. He believes that certain problems in biology, such as how the brain works, are intrinsically insoluble and cannot ultimately be reduced to explanations at the molecular level. He also does not believe—contrary to the accepted view—that the DNA contained within the egg provides the programme for the development of the embryo. He argues that the genetic material is just one of the components in a complex interacting system, and sees embryonic development in similar terms to that of a coral island: the result of a large number of predictable interactions that nevertheless give rise to a quite ordered structure. In his book *Paradoxes of Progress,* he argues that science is running out: that soon all the problems capable of solution will be solved.

His pursuit of this argument has led him into the field of hermeneutics, the theory and study of how we should interpret such things as the scriptures or the findings of science. He now supports his thesis with the contention that the paradigms of science are heavily influenced by the essentially theological traditions in which they operate. In this analysis the science of the west is a very different activity to the science of the east, and reductionism is, at least in part, a subjective, cultural phenomenon.

In a similarly heretical vein, Stent has also argued that style in science is as important in determining outcome as content. If, he suggests, Watson and Crick had not discovered the structure of DNA the way they did, or been the kinds of personality they were, and the information had, instead, emerged in a cautious, piecemeal fashion, the impact would have been very much less and the course of science altered.

Much of Stent's work in molecular biology was done with Max Delbrück's phage group at the California Institute of Technology. Delbrück's influence, not just on Stent, but on the whole field of molecular biology has been profound. Phage are viruses—little more than sequences of genetic material—which invade bacteria, and direct their unfortunate hosts to begin synthesizing viral proteins. Delbrück, who had initially trained as a physicist specializing in quantum mechanics, was one of the first theoretical physicists to see the possibilities opening up in biology. He recognized that the simplicity of phage made them an ideal system for studying the action of genes at the molecular level. He described them as 'a fine playground for serious children who ask ambitious questions', and it was using this system that he was able to establish that DNA was the genetic material. Each summer in the 1940s, he ran an enormously influential course on phage at the Cold Spring Harbor Laboratories outside New York. Among those it attracted were people such as Jim Watson and Seymour Benzer who were to shape the new field of molecular biology. Together with his collaborators Salvador Luria and Alfred Hershey he received a Nobel Prize in 1969.

Stent, like Brenner, now works on higher organisms. He studies the development of the leech, and in particular the nervous system. Given what seem to be his rather gloomy views about science, one of the first things I wanted to ask him was why he did it all. I was also curious to know how his thoughts on the way science operates would affect his view of a figure such as Max Delbrück. And as Stent, like Delbrück, Crick, and a number of other molecular biologists, began his career in the physical sciences, I wanted to find out what made him change direction at what was to be exactly the right time.

––––––––

When I was a teenager my main interest was railways. I was living in Germany, in Berlin, and all my free time was spent at the Berlin railway stations watching the trains go by. They were my main interest in life. So my plan was to become either something like a railway engineer or a civil engineer. Then I emigrated to Chicago, finished High School there, and went on to University. I decided to study engineering. But somehow, for reasons that I can no longer recall, I signed up for chemical engineering, not for civil engineering, so I studied chemical engineering for one year at the University of Illinois. But I didn't like the drawing.

We had to do engineering drawing, and I was very poor at that. So I decided to get out of engineering and I went into chemistry. So in my second year at the University I became a major in chemistry. Then I found I disliked organic chemistry very much. But I did like calculations, and somebody told me that *physical* chemistry was where you did calculations. So I became a physical chemist and I finally got my degree. My undergraduate degree is in essentially physical chemistry.

'Had you any idea what you were going to do with it then?'

Well, not exactly. I knew that there was such a thing as Graduate School, but this was all very vague in my mind. It was during the war, and I had a job during my senior year on the synthetic rubber research programme, which was then run by the Office of War Production at Illinois. They offered me the opportunity to do a Ph.D. in physical chemistry while working on rubber and that's what I did. So I became a high polymer chemist working on the physical chemistry of rubber, polymerization and so on.

'You're not in biology yet!'

No, no. I got into biology after my Ph.D. There was a woman that I knew at Illinois, Martha Baylor, who was one of the earliest American electron-microscopists. They actually had an electron microscope! This was in the early forties. She knew some biology because she had taken Delbruck's first phage course at Cold Spring Harbor in 1945. She gave me Schrödinger's book *What is Life?* to read, and said, 'He's talking about this man I know. He's talking about Delbrück'. And I was absolutely fascinated by this book, and I began to formulate a plan because I was very bored with physical chemistry.

'Why didn't you like it?'

Because the spirit of research, at least at Illinois, was that all the important things in physical chemistry had been done at the turn of the century and you had to discover something left that hadn't been done yet. That was the art of research! Then you had to do it, publish it, and then find something else that hadn't been done. And, although at the time I couldn't formulate it precisely, I found this very unsatisfactory. Then I read about the gene! I had never taken a course in biology, and the only time I had actually heard about genes was as an undergraduate when I took a philosophy course. It was on the great minds of modern times and was about Freud, Darwin, Marx, and Hume. And it was in the course on Darwin that I for the first time heard the word 'gene'. So, my acquaintance with genetics was completely confined to this one philosophy course. Anyway, I left Illinois for a year to go to Germany to work for the military government, and while I was there I reflected more and decided that I didn't want to stay in chemistry. So I went back to Illinois to finish my Ph.D. on rubber and during that year I wrote to Delbrück asking him to take me on. Delbrück, when I first heard of him, had been in Tennessee and I wasn't too keen to go to Tennessee. But when I came back from Germany, in 1947, my friend Martha told me that he had become a Professor at Caltech. My dream had been to go to California in any case, so I decided I'd like to go to work with him. So I wrote 'Dear Professor Delbrück, I want to go into biophysics, do you have a place for someone like me?' He sent me back a postcard—Delbrück then only corresponded with postcards—and the answer was 'No, I don't have any place for anybody like you.' So that ended my career in biology at the time.

'That's quite a devastating reply. Were you devastated?'

I was completely crushed, yes. Because it had already taken a lot of nerve to write to him in the first place. But then a friend of mine said that the Merck chemical company had just established a new kind of fellowship, a post-doctoral fellowship. Now, post-doctoral fellowships like this were very unusual at that time, and this was specifically for people who had been trained in chemistry to go into biology. So this friend of mine told me 'That's what you want to do. Why don't you apply?' So why not? I applied for the fellowship, saying that I wanted to go and work with Delbrück. And about six months later I suddenly got a telegram: please come to New York next week, for an interview. That was the first time I took a trip at somebody else's expense, my first taste of free travel! But this interview was absolutely devastating. It was terrible. They asked me: 'So, you want to go into biology. What do you want to do?' I said I wanted to test whether the second law of thermodynamics applies to living systems. They were all shaking their heads and after the interview I was dismissed summarily. I went back to Illinois totally depressed. Two days later I got a telegram saying that I'd got the fellowship. When I got this telegram I wrote to Delbrück and said,

'Hello, you probably don't remember me any more, but I have this fellowship to come and work with you.' And Delbrück replied that yes, that was fine and so forth, and said 'Undoubtedly you want to work on phage.' So I said, 'Yes, I want to work on phage.' I didn't even know what it was. So, my life changed completely after that, because I had no more contact with any of the people I knew before. It was like being reborn, like being a born-again Christian. I was 24.

'Did your family or anyone disapprove? What did your friends feel about this change?'

None of my family were academics. My father thought chemistry, biology, the whole thing was a waste of time anyway—I should be in business! So it seemed quite irrelevant what I was doing. But I think most of my friends at Illinois thought I was out of my mind too, because the usual thing for all my colleagues was to go to work for Dupont or Monsanto. Most of the chemical industry in the United States was then run by graduates of this chemistry department at Illinois. It was immensely powerful. Very few people went into teaching and to go into *biology* was considered completely nutty.

'So it was very exciting, being in this new environment?'

Yes. Delbrück told me 'You're going to take the phage course at Cold Spring Harbor.' So I said 'Yes sir. I'll certainly do that.' And there it was immediately obvious that research was something entirely different. There were a million problems, and the art of research was to pick which problems from the million of unsolved ones, rather than having to find something that hadn't been done yet. So this was like being let loose in a candy store where you can eat all you like. And that summer at Cold Spring Harbor people like Jim Watson and Seymour Benzer were my classmates. These people were all of a quality of individual totally unlike anything I'd known at Illinois.

'Can I just ask you about Delbrück? Do you think important figures, like Delbrück, really shaped the whole development of molecular biology?'

Oh yes. Delbrück's influence in molecular biology is very curious in that he was a kind of Gandhi figure really. His strength was his incorruptibility. He was not a leader in the sense that he actually made good suggestions or that his intuition was especially good. It wasn't bad, but his role was essentially *moral*. Everyone's intention or aim was to please him. Prizes, or recognition by others, were only secondary. The main thing was, when you did something, you were hoping that Max would approve of it. And therefore things like stealing or cut-throat competition didn't exist because Max would see through them. It would be like God, you see. If you did something illegal, maybe you'd get away with it with a fellow mortal, but God would know that you'd cheated.

'And did everyone feel the same way about Delbrück?'

All of the group, the so-called phage group, felt that way. So it created order. Now the point was that very often Max's opinions were incorrect, and so it was also a test. You often had to do something against his views, but he would respect it. You would say 'I want to do this', and he'd say 'It's nonsense. It will never show anything.' But one did it anyway, and if you could then show him good results, then he would of course immediately honour them. He would feel even better that against his advice, you had prevailed. But he was always the standard of integrity, and that is what made this movement possible. And the people whom he thoroughly disapproved of were just out. Of course, if the leader is bad then it could be bad for the field. It's not without risks. But I think it's probably essential that there is a leader. And so the field that you and I are interested in, development, is partly in a bad state because there is no ideology, no leader who is setting the tone, who is some kind of court of appeal.

'You were now within the court, as it were, of Max Delbrück, and you found science satisfying in a way you hadn't found before?'

I would say I already found science itself satisfying at Illinois, even as a physical chemist. Because what I liked most was being 'a scientist'. My interest, strangely enough, in science itself, is not all that deep. My main interest when I got into science was to be a scientist. I liked that as a lifestyle.

'I want to know what that means.'

Well, the line of work—to be in a lab. To work in a lab is to have friends with whom you discuss problems of mutual interest. Then you discover that you can travel as a scientist, and go all over the world. You have friends everywhere. So, I like the social aspects. I consider that one of the tremendous satisfactions of being in science. And when I find something—the rare times you do find something new, which isn't very often—I would say my main satisfaction in discovery is the thought that next time I go to a meeting I really will have something to say.

'It's not an intellectual gratification then?'

No, the intellectual gratification is much less than the expected reward that I'll have when I see my buddies the next time, and tell them 'Look here man, I found this!' This is what I like best about science. I'm always thinking about the papers you see. Even before I have found something, I'm already thinking of the opening phrase of the paper in which I will describe this discovery.

'You're already thinking of the title, and the opening phrase?'

Yes, before I've found anything, yes.

'And publishing the paper gives you great pleasure?'

That's right, yes. And, you see, many of my collaborators, although they're excellent men, they've hardly published. It's always a tremendous struggle to get them to publish papers and so on. I can't understand that. So there's some kind of gap between my dear collaborators and myself. Because if I didn't push them, they wouldn't publish for years.

'And the gratification of publishing is really so that other people will see your work—it's sort of an ego-trip?'

I don't want to deny that. The expectation of admiration probably is there too, but I think it's not only that. I think it is just the pleasure. I like conversation very much, and to make conversation you have to have something to say. And so it's partly that, I think.

'And so the paper is almost a way of initiating the conversation?'

That's right, yes. It's a bit like a novelist. I imagine with a novelist there's probably an ego-trip, that he would like to be famous and have his novel admired. But I think that's not all of it. He has some urge to tell what he knows.

'But what about discovery? You've said that the gratification really lies in having something interesting and new to tell people, but isn't the actual solving—the moment of solving the problem—not an exciting moment for you?'

Of course it's exciting, but I think the final excitement, the real source of the gratification, is not so much beating nature as being able to tell it. Of course, if I have a theory, 99 per cent of the time the theory is wrong, but that 1 per cent of the time when I have a theory, and I get experiments that confirm it, then of course I'm very happy for it to turn out that way. But I think the happiness mainly derives from being able then to write a nice paper. Let's say, if I was on a desert island, Robinson Crusoe, I think I wouldn't do science.

'Because there'd be no one to tell it to?'

Exactly. If I was marooned on an island with a lab, I don't think I'd do any experiments.

'That's a very nice way of putting it. Now, I know that you're interested in the philosophy of science. Do you think there's something called scientific method?'

Not that you can formulate. It's not the way I think. One can make some generalizations, but it's more an a posteriori justification . . . .

'So how do you actually go about your own science? It's a widely held view, for example, amongst people who are not scientists—and even some scientists believe it—that the way to go about science is to have an absolutely open mind.'

No, that's complete nonsense. You can't. There's no such thing as an open mind. First of all, it's a psychological impossibility. But even if it were true, then you would be condemned to inactivity, so there's no such thing as an open mind. You have to have some kind of prejudices and approach the world with some kind of theoretical framework.

'I'm so glad to hear you say that. Scientists are very prejudiced and, it seems to me, that is what actually gives the dynamism to science. That's what scientific imagination is.'

Yes. The picture of the scientist as a man with an open mind, someone who weighs the evidence for and against, is a lot of baloney. Scientists are studying the world, the outer world, and that presents us with an infinitude of phenomena, so you could not possibly address all things. You have to make abstractions. You must select a sub-set of phenomena to attend to. And this selection must, by necessity, be theory-guided. So by the very fact that you focus your attention on just some limited thing, you're prejudiced from the very beginning. This open mind business is a lot of nonsense.

'Do you think, then, that there is any analogy between art and science? There's a lot of slightly romantic discussion about the creativity of scientists being the same as the creativity of artists.'

I think there is some similarity. There's also a difference between art and science. The world that the artists address is the inner world, so the fundamental difference between a scientist and an artist is that the artists address the inner world of the emotions, whereas the scientists address the outer world of physical phenomena. But I think the similarity is in the act of discovery. I don't believe that art is only interested in entertainment. It is similar to science in that it endeavours to discover truth. The artist endeavours to discover truth about the emotions, the inner world. It's not a question of tests or proof, but of validity, whether the experience seems valid to you or not. It's a subjective judgement. If you feel that reading Dostoevsky provides you with some insights, then either you feel that or you don't. There's no way of proving it, and there's actually no need to prove it.

'But is the process of discovery in art and science not very different?'

I don't think it's so different. If on the one hand you go to physics classes wondering 'Why does the sun rise, and go down?' and they tell you about gravity, and Galileo and Newton, then you feel good. At last you understand what's going on up there, right? I think this is not so different from when you read Dostoevsky let's say, or when you gain insights from Shakespeare. You understand about people, what makes them do what they do. It enlarges your understanding of the world. So I think, in that sense, the act of discovery is not so different. But it's a different phase of the world.

'But you're talking about responding to science and responding to art. What about the actual process of doing art or science?'

That is very different, but the end result is still this psychological gratification.

'Is that why you've moved into this new field of hermeneutics?'

I have developed a side interest in philosophy. I'd already studied it as a student, and I was interested then, but my enthusiasm was really awakened when I worked in Japan. I was there in 1960 on a sabbatical. In the labs, there were people in white coats working with ultra-centrifuges, phage, mutations and so forth, just like anywhere else. But I was amazed to discover that they were actually fundamentally different in their approach. Their view of what they were doing seemed to be quite different from mine. So, my interest in epistemology and philosophy was aroused by this personal experience of the radical difference between Japanese science and what I knew to be Western science.

'So what was the difference?'

It has to do with the notion of reality and truth. When I wanted to please Delbrück, I wanted to publish something that was true, since only that would please him. But what I found in Japan was a much more aesthetic notion. The element of truth was not paramount. For them, writing beautiful papers was very important, and the beauty of the paper was paramount rather than the truth. I first got on to this when I realized that the notion of a control experiment seemed to be foreign to the Japanese. They're mostly positive thinkers. To them a control experiment is negativism you see, like trying to tear things down. They don't like that. Also, during seminars, the type of questions that were asked would never be critical questions. At first I thought it was just politeness, because Japanese are very polite. But it actually has a much deeper philosophical and religious basis. It's Buddhism as opposed to Christianity. I think Western science depends on the notion of law and order. Historically, you can trace the development of this notion from that of an orderly universe created by God, who made the laws. Moreover, He created us in His image and therefore it is given to us to divine the reasons that He, in His infinite wisdom, had in designing the world. And so the whole enterprise of science, metaphysically—I'm speaking about the metaphysical basis of Western science—depends on this credence: God, the Creator, made the laws, created us in His image, and therefore we dig for what His notions are. There's some chance of finding out, you see. Whereas, for the Buddhists, this concept is considered to be the height of naivety. Because, for them, anybody who has any sense knows that the world is infinitely complex. So that's the radical difference. Because if you believe that there are no laws, and that there is no orderliness, therefore there is also no truth. Thus it's all a question of subjectivity which is exactly what avant garde philosophers of science are saying now. But the Japanese people have felt this for centuries, for millennia, in their heart of hearts.

'But I've always believed in orderliness, simply because it's apparent to me that the world is orderly. I've always seen patterns in the world.'

Regularity, of course, is a fact of experience which we see, and that we get in the cradle. But that is different from believing that this orderliness is, in fact, out there. We're taught that orderliness is a reflection of underlying laws which it is given to us to discover. That is not necessarily the same thing. That I think is somehow a reflection of a deistic belief. And I do believe that while the Buddhists were wrong in the short run, because it turns out that the world is more comprehensible than they thought, in the long run they were right. Now that we've pushed science to its limits, we see that they were right after all.

'Are you then not a reductionist? Do you not believe that all human behaviour can be simply reduced to molecular biology in the long run?'

On the contrary, I believe that science is, by nature, reductionist, but I also believe that reductionism will not carry us all the way. One of the reasons why I think science will eventually peter out is because you must always explain some higher level in terms of some lower level—that's what scientists have to do. But I think that when finally we get to sufficiently complex things, this will not be possible. It is precisely because I think reductionism will have to fail, that I believe that science is coming to an end.

EVOLVING
IDEAS

JOHN MAYNARD SMITH *was born in 1920 and is*
Emeritus Professor at the
University of Sussex.

CHAPTER 10

MAKING IT FORMAL

JOHN MAYNARD SMITH

Evolution theorist

IT IS very hard to believe that all the marvellous forms of life—octopuses, snakes, eagles, human beings—have evolved through a series of chance events. Yet this is the essence of Darwin's theory of natural selection. It is chance which dictates the particular genetic mix of any individual in a population, and chance which determines whether any aspect of that mix happens to put the individual at an advantage over others in the face of environmental pressure. As those with some advantage in the competition for food, space, and so on, will clearly survive better than those without, natural selection ensures that the better adapted variants are passed on to the next generation. As the balance within a population shifts in this way, new species evolve. It is a scenario which rejects the idea that we can inherit from our parents traits that they have acquired in their lifetime. Giraffes do not have long necks because their ancestors were always straining towards higher branches—rather giraffes with longer necks survived better because they could reach the leaves on the higher branches.

It is an account of our origins which, even when not the subject of controversy on religious grounds, seems to produce considerable unease. Certainly many people find it counter-intuitive. And in recent years the picture has also, perhaps, been confused by the advent of sociobiology. As evolutionary theory has become concerned not just with why animals have come to look the way they do, but also with why they *behave* the way they do, it has become seductive to ask to what extent human behaviour is directed by our genetic inheritance. If altruism or aggression in animals has a genetic basis, why not in humans? It is not difficult to see why people who find such speculation disturbing might wish that evolution itself had happened otherwise. Nevertheless, even away from the cloudy waters of sociobiology, a deep suspicion of evolutionary theory remains. What, I wondered, did evolutionary biologists make of all the discomfort surrounding their chosen field?

Maynard Smith is a distinguished and outspoken evolutionary biologist who has made important contributions in a number of areas. In particular he is known for his work on the evolution of sexual reproduction, and for his influential theory of the evolutionarily stable strategy. Originally developed to explain why male animals of the same species engage in ritual combat at the time of mating, instead of actually damaging each other, it is now recognized as a general principle in the evolution of animal behaviour. Did he ever, in his darkest moments entertain any doubts about the foundations of evolutionary theory? Why did he think it is still so often the subject of public debate?

As we walked across the campus of Sussex University where he is Professor of Biology we began by talking about Maynard Smith's own introduction to the ideas of evolution. He is yet another biologist who first trained in the physical sciences and it was not until after the war that he transferred his mathematical skills from the problems of designing aeroplanes to those of evolutionary biology. His mentor then was the intellectually formidable J. B. S. Haldane, who was Professor of Genetics at University College, London. A mathematician and classicist turned biologist and popularizer, Haldane had shown how Mendel's genetics provided the mechanisms of inheritance that natural selection required, and helped to put evolutionary theory on a firm mathematical basis. Haldane had influenced Maynard Smith since he was a boy at school, so how was it that engineering intervened?

———

I didn't actually learn any science formally until I was, I think, over 30, but I've always been interested in science and I've always read books about animals and about physics and so on. That was on my own account . . . my school didn't teach that sort of stuff.

'What do you mean, your school didn't teach you that sort of stuff?'

Well, my school was a ridiculous place called Eton which in my day, at least, taught mathematics and the classics but basically not science. So I didn't do any form of science at school or at university.

'So how did you come across it? Was it through your family?'

Well, let's give Eton its due, there were some books about science in the

school library, if you went there. I don't know, really, how I became
interested in science. I was taken to places like the Natural History
Museum when I was quite small, six, seven, or eight, and I have been
interested in animals ever since I can remember. How I ever became
interested in physics and chemistry, I really can't imagine, because
certainly no member of my family knew what an atom was, I don't
think. I mean, they were deeply ignorant people. They were landed
gentry who hunted foxes and things like that, but they didn't think very
much. So I really don't know how I became interested in physics but I
certainly did, and from the age of about thirteen or fourteen on I read
books about physics and chemistry.

'Almost secretly at school?'

I don't remember actually having to be secret about it, but I do
remember—and I think it was partly because I wore glasses from the age
of about three—being called 'professor' by my fellow schoolboys, even
at my prep school. So I must have been that sort of idiot even then, I
suppose. But I don't remember being secretive about it particularly.

'So how then, with this sort of education, did you end up becoming an
engineer?'

The family had a stockbroking business. The idea was that I was going to
enter stockbroking and become very rich—my grandfather was head of
the firm—and I realized there was really no way I could possibly do that.
I remember announcing at Sunday lunch, which was a big family
occasion, that I had decided I wasn't going to go into the City, and my
grandfather, who was a lovely old chap, said 'Well boy, what *are* you
going to do?' This took me by surprise because I hadn't really thought
that out. But several weeks earlier I'd been to a lecture by a man who
directed the building of the Sydney Harbour Bridge. It was terribly
exciting getting the two halves to meet—they had these male and female
fittings that had to slot into one another. So more or less off the top of
my head I said, 'I'm going to be an engineer', visualizing myself erecting
great bridges. Once you'd said something to my grandfather you
couldn't go back on it, so I was stuck with it. So I went and became an
engineer and it worked out quite well actually.

'You became an aeronautical engineer, not a man dealing with bridges.'

By the time I graduated in engineering, there was a war on and they didn't want bridges; they wanted aeroplanes. So I went away and designed aeroplanes for five or six years.

'And you enjoyed that?'

Yes, I did. I would probably be doing it still if, again, my eyesight hadn't made a difference. I can't fly aeroplanes because I can't see well enough. There's something slightly emasculated about designing something you can't fly. It's rather second-rate it seemed to me, and I never really grew to love aeroplanes in consequence. I actually enjoyed the purely technical problem of solving the question of how to make the thing fly, how to make the wings light and so on, but it was purely a technical problem with no real emotion attached to it. I wasn't really all that interested in aeroplanes as such . . . they're noisy and dangerous things.

'And so you tried to get out of engineering. Was that a conscious decision to move on to something else?'

Oh yes. To be honest, after the war, the firm I was working for did go broke. But I had decided to leave aircraft engineering before that, and had quite a difficulty making up my mind whether to go in to physics or biology.

'How did you decide?'

Well, I decided that physics is too hard. It really is. I'll never make an experimentalist; any piece of apparatus that I come within miles of fuses or blows up or sets light or something. So I was only going to be a theoretical physicist. And you must remember, I still didn't know much physics. I knew about Einstein and Newton and other people from my own reading, but there was no way I was ever going to do what Einstein or Newton did, I knew that. It was clear that physics is just too hard. Biology is actually quite easy, as you know. So I thought, well, I've always loved animals and it seemed a perfect way of putting the two things together.

'So you chose then to go to University College. Was that because Haldane was there?'

Yes. Whether I would have chosen the same way if I'd known about him what I do now, I don't know. Not that I have anything other than affection and respect for him, but he was actually rather a dangerous person to live close to.

'In what way was he dangerous?'

Well, he was about 16 stone and very irascible. It's difficult to explain. He was just very large, both physically and intellectually. I mean, he's the only person I've mixed with continuously who was manifestly cleverer than I am. I'm fairly arrogant about this but he really was very, very clever indeed. So he made you feel stupid. He also made you feel very small and he had a very short temper, and it was rather like sitting on a landmine which was liable to blow up at any minute. I saw him so often with other people . . . he'd meet someone and he would be absolutely charming to them until suddenly he would realize they didn't know what they were talking about and then you would see him sort of switch off, almost like a light going out. I worked with him, I suppose, for 10 years, waiting for the moment when he would realize I was stupid. And he never quite did, but I was constantly believing that one day I'd see that expression go across his face and . . . it was frightening.

'So you lived in constant fear?'

Yes, absolutely so, and indeed all his associates I think in some sense did. Much as we loved him. And we did love him.

'What was the lovable quality about him then?'

He was immensely kind and generous, so long as you didn't pretend to be somebody you weren't. He was always exceedingly good to me.

'And did he have a big influence on you intellectually?'

Oh yes, long before I even met him. I do remember, even at Eton, discovering that there was one person above all else whom my teachers hated, and this was J. B. S. Haldane, because he was against the monarchy, the establishment, and the established religion and the rest of it. But also he was a renegade because he was one of them—he'd also been to Eton. I remember thinking that anybody they hate that much can't be all bad. So I went off to the school library to see whether there were any books by Haldane, and to their credit, I suppose, there were. I found *Possible Worlds,* and I can still remember the excitement of reading it. The mixture of intellect and blasphemy was absolutely overwhelming and I've been attracted to that all the rest of my life. It was one of the moments which I really still remember.

'Did Haldane influence a whole generation in the way that he affected you?'

Yes, I think he did. I think he influenced a generation more through his *writing* than through personal contact. When you come down to the people who actually worked with him, there are relatively few of us— because basically you had to have a certain character of courage to benefit from working with Haldane. But when it was slightly diluted, when contact was in the form of essays and so on, you got the excitement without the terror. Yes, I think a whole generation *was* influenced by his way of thinking. It was a combination of appearing to know everything—in fact, there were many things he didn't know— that he'd read everything, that he understood everything, and yet he was so simple and direct that he gave the impression that you could be like that. He always made things very, very simple to understand. Haldane's mathematics is so transparent that a child of sixteen could understand it. He gives simple direct illustrations of everything he says. There are no long words in Haldane, it's all very simple, factual, clear, and yet it has this extraordinary breadth. I think great scientists do have this gift of seeing the world as simple, and Haldane had that gift.

'But it's something I suppose that's quite hard to pass on?'

Well, you can imitate. I mean, I've spent my life imitating Haldane.

'Consciously?'

Well, consciously and unconsciously, and for that matter, imitating Peter Medawar who is the other great figure who influenced me as a scientist. I think in the case of imitating Haldane, it's been largely unconscious.

'Was it through Haldane that you came to think about evolution?'

No, I'd been interested in evolution a bit earlier than that. I had religious doubts from the age of about ten or eleven. I remember how I actually first got religious doubts. I lived in a little village called Exford in the middle of Exmoor. Stag hunting was the industry and the religion of Exmoor, and it was a very charming fact about the village in those days—it may still be true for all I know—that on the first Sunday of the stag hunting season we always sang as the first hymn in church, 'As pants the hart for cooling streams when heated in the chase, so pants my heart oh, Lord for thee,' and so on. So I grew up, therefore, thinking that stag hunting and Christianity were basically all one thing. Then the old parson died and they got another young man from up country somewhere and he was against stag hunting, and the first hymn in church wasn't 'As pants the hart for cooling streams.' I remember feeling very disturbed by this, and my grandfather was furious, you know, stuttering with anger, and I began to wonder whether he was right. I remember going to my mother and asking her to point out the place in the Bible where Christ went stag hunting. I must have been quite young, but obviously I was beginning to ask questions, and by the time I was thirteen or fourteen my doubts were pretty severe, and they were quite largely fuelled, I think, by evolution. I already knew about dinosaurs and about Darwin and so on, and I think my original interest in evolution was really almost a philosophical one; how did we get here, were we really created, and ultimately, whether religion as I was taught it was true, which by and large, I think it isn't.

'When you went in to biology, were you really thinking that there were major problems in evolution that you wished to solve? In a way, one might have thought that evolution wasn't a terribly intellectually exciting discipline to work in, that the major problems had, perhaps, already been solved. Did you see it like that?'

No, I don't think I did. I think what you say is true to this extent, that in 1950 or thereabouts, when I started studying biology, I did think that the problems of evolution had been by and large solved, and that what I would do was to learn about the solution. I wasn't really thinking about what contributions I might make, and indeed, for the first ten or fifteen years after graduating as a scientist I wasn't primarily working on evolution. As a matter of fact I was working on problems of development, as you do; I was working on problems of ageing, and on certain technical problems in genetics to do with the effect of inbreeding and so on. But I wouldn't really have thought of myself as an evolutionary biologist and I suppose it was really only in the early sixties, when it became apparent to me that there were actually vast confusions in my mind and in other people's minds about a number of evolutionary problems, that I really started thinking very seriously about evolution.

'Is that how you came to your famous evolutionarily stable strategy?'

Yes, I think ultimately it was. I should explain that when I was a student, in 1951–2, I knew about the work of the ethologists, Tinbergen and Lorenz, and in particular I was familiar with their idea that animal fighting was often very ritualized, that animals didn't bite and scratch and kick, but they displayed at one another. I didn't doubt the fact, but even then I knew there was something wrong with the explanation being offered by the ethologists. But as it happened I didn't work on it for another twenty years. It was just there at the back of my mind as something that clearly needed work but I hadn't got round to it. The problem basically was that the explanation being offered was that animals don't like to scratch and kick and so on because it would be bad for the species if they did. I already knew enough about genetics and the evolution theory to know that natural selection isn't there for the good of the species. It's, so to speak, there for the good of the individual, and what I knew was needed was an explanation in terms of the individual.

'So how did you arrive at the solution?'

Well, rather accidentally. It arose basically because I was sent a paper by the journal *Nature* to referee. It was by a man I'd never heard of called George Price, and this contained what looked like an explanation, but it wasn't worked out, it wasn't made mathematical. Nevertheless it seemed all right and I recommended that it be published. I thought no more about it for a while and then I thought, well, let's see whether I can make it mathematical, whether I can make it rigorous.

'Did *Nature* publish the paper?'

No they didn't. The manuscript was some sixty pages of typescript and they clearly couldn't publish it as it stood. I suggested that they ask him to give them a short version and he never did. But after I'd done the mathematics, I tried to get in touch with this guy because I wanted to quote his paper and it didn't seem to have appeared in *Nature*. He turned out to be an amateur. He was living all by himself down Charlotte Street, just round the corner from University College. He was an American but he'd come to live in and work in England because he wanted to work on theoretical biology, and he thought he could live on his savings longer in England than in America, which in those days you could. He was apparently wholly unaware that you could actually get grants to support scientific research, and as a matter of fact it isn't all that easy to get grants to support theoretical work. So I got together with him, and after a great deal of struggle we managed to write a joint paper in which we put his original idea and my algebra together.

'What was his original idea?'

Well, the original idea was really horribly simple, I'm afraid, as all ideas are. It simply was that if you and I are having a fight, the reason why I don't hit you is that you might hit back. I know that sounds incredibly naive, but basically the idea is that animals are ritual in their behaviour because of the risk of retaliation. What I had to try to show was that evolution could produce a stable state based on the risk of retaliation, and that isn't actually all that easy when you try to make it formal.

'Putting it into mathematical terms, in other words. Was it really quite difficult?'

Yes. Again, the algebra wasn't all that hard but the difficult thing in applying mathematics to science is to, so to speak, turn the real world into algebra. That's what I learned from Haldane I suppose. It's the thing at which he was the supreme genius. The problem was to somehow or other take this very qualitative verbal idea of Price and turn it into something which you could formalize mathematically and, therefore, calculate the consequences.

'It has been a very fruitful idea, the evolutionarily stable strategy, hasn't it, because it makes sense of all sorts of animal behaviour?'

I think that is true, but let me hasten to say that the idea, as with most ideas that turn out to be fruitful, wasn't just mine. A whole series of different people had what essentially is the same idea independently, but in a different context. The basic idea is to look for a strategy—that is, simply a behaviour or a structure—which is stable in the very simple sense that if every member of a population does it, then there's no other strategy that somebody could come up with which would do better. So you search around for a strategy which is literally uninvadeable. And it works for everything from what sex ratio of kids you ought to produce to whether you ought to migrate or not.

'Is that what you would call sociobiology?

'Well, sociobiology is usually thought of as the study of the social behaviour of animals, including man. The idea of stable strategies is relevant whenever the best thing for an animal to do depends on what all the other guys are doing and, of course, that's always true in our society. So, certainly the idea is very much bound up with the study of society. Let me hasten to add that I don't actually like the term 'sociobiology'.

'Why not?'

I'm very interested in evolution of social behaviour of animals. I think that human beings are actually so different from other animals in the degree of cultural and ethical and mystical and religious and political concepts which influence their behaviour that it isn't wildly fruitful to think about them just as if they were another animal. I think that what Ed Wilson has done for us by introducing the term 'sociobiology' is to make it harder to think clearly about animal behaviour and not easier to think about human behaviour. And I suppose I'm really showing another aspect of my upbringing. I was a young man when Hitler was in power, I was in Berlin in 1938 just leading up to the Munich Settlement, and the whole of my thinking about the world has been much influenced by belonging to that generation. For me, the applications of biology to human beings means Rosenberg and the race theories, so I'm obviously a bit reluctant to get involved in biological applications to human behaviour.

'It's not that you don't feel that it's possible, you just think one has to be extremely circumspect about it?'

I certainly don't think its impossible. But I do think it ought to be done by people whose primary subject and expertise is a knowledge of man and society, not by people whose primary expertise is a knowledge of ants or baboons.

'I take your point. Can we come back to evolution itself. You spoke about it being in conflict with your religious beliefs and that was obviously some sort of small struggle for you. It does seem to me that it is a struggle for a lot of people. Evolution does arouse a great deal of passion. Do you think it is because it undermines religious beliefs that people do get so het up?'

Yes, I'm sure it is. Even when people are not, so to speak, overtly religious. The basic point is this. Every society and every group of people that we've ever known has some kind of myth about origins and where they came from. This is usually seen as part of their religion, like the story in Genesis. The object of this myth is to place man in the cosmos and in society and say why he's valuable and how God minds about him, and so on. Darwinism is also a story about origins—it says where man comes from—but it has a totally different function. It's not actually intended to tell you that man is particularly marvellous or that God loves him or anything like that. It's just to tell you what happened. So these two things are bound to come into conflict. It's surprising, and I could document it if you like, but it turns out that the socialists, the women's movement, and the gay liberation movement have all in recent years attacked Darwinism because it doesn't give the right image for their particular view of man. And I think they're just simply wrong. I don't think Darwinism is trying to give an image which justifies either socialism or capitalism or homosexuality or heterosexuality. It's just an account of what happened.

'But you feel that these groups are looking for an origin which will justify their own position?'

Yes. It's made most clear, as a matter of fact, in Bernard Shaw, and in the preface to *Back to Methuselah*. In *Back to Methuselah* Shaw, quite explicitly, wrote a play to produce a theory of evolution which would have good moral effects rather than the bad moral effects of Darwinism. He makes it perfectly clear that that's why he's writing the play, and that he thinks Darwinism has all sorts of unpleasant moral implications and, therefore, has to be wrong. Now Darwinism is right or wrong quite independent of any moral implications it may have. We don't decide whether an idea is true by whether it has good consequences.

'Do you think the other reason, apart from the moral one, is that Darwinism is really counter-intuitive? The idea of selection working on random change seems such an inherently unlikely way, as it were, to go from the amoeba to man'.'

I find it hard to say whether it's counter-intuitive because I suppose Darwinism was one of the first significant ideas that I mastered. I must have mastered it at the age of perhaps thirteen or fourteen, and it has been so much a part of my thinking for so long that people who find it counter-intuitive seem to me to be just blind.

'Well, I must confess that I find it counter-intuitive. Do you really never have doubts that there is not something else other than, as it were, crude Darwinism? Are there no dark moments when you think that the evolution of the middle ear or the eye or something as complex as that may involve some other mechanism?'

No, I don't honestly think that I have doubts. I have doubts of rather a different kind, technical doubts. I'm aware, for example, of experimental results or specific observations which seem to cast doubt on the genetical theory that I hold to be true. Every theory has experimental results which suggest it's false and I certainly have doubts about those things. I have doubts about the principle that acquired characters are not inherited. I think on the whole that acquired characters are *not* inherited, but every now and then some fact is reported to me which gives me doubts, to wonder whether in fact I'm right about that. But no, I don't really get doubts when I look at the eye or the ear. I grant you that it's complex, but the problem is really, oddly enough, a problem for you and not for me. You're a developmental biologist and I'm an evolution theorist. If you do the sums, there's been plenty of time for natural selection to, so to speak, programme all the DNA that is in your genome and mine, much more time than necessary to get that DNA in any order that selection wanted it. So the problem is how the DNA programmes your own development and that's a deep problem, as you know.

'Hasn't Popper argued that evolution is not really respectable science, that it's all some rather boring tautology and is not to be taken terribly seriously?'

Well, I think Popper himself would (a) claim that he never said it, and (b) claim that if he did say it, he didn't mean it. He's been backsliding on that view recently. But there are two things one could say. First of all, the general proposition that evolution has occurred by natural selection is a perfectly falsifiable theory. It would be falsified, as a matter of fact, if acquired characters were inherited, because if acquired characters were inherited then that would be another mechanism of evolution not accounted for in our theory. But, of course, most of the time one is not trying to falsify or study the theory of evolution as such, but the theory of evolution of some specific trait, such as the evolution of sex or something. There you can form different hypotheses and test them in just the way that Popper would approve of.

'I do sometimes get the feeling that the national press are really creationists at heart. I don't feel that the great British public really believes in evolution—or would love not to believe in evolution.'

Well, I agree with you entirely. For example, the paper I take, for better or worse, is the *Guardian*, which has perfectly respectable and sensible science correspondents. It also, or so I deduce, has a little man they keep in a box, and they let him out once every three months to write an antievolution editorial and then put him back in the box again. I've no idea who he is, but I read his editorials with fascination.

'It is a curious phenomenon; I know what you mean.'
Do you know who he is?

'No, I don't know who he is, but I must say, I'm not sure if he doesn't pop around to other newspapers too,'
Ah, well, I don't see other newspapers. Maybe he does the same thing for *The Times* on off days.

'Has it been fun working on evolution?'
Oh yes.

'What's been the pleasure?'
Well, there are three levels of pleasure. First of all, I love looking at animals; any excuse to go out and watch animals doing things is a pleasure. Secondly, I love doing sums and the actual pleasure of getting a sum to come out is very great. I know that people who aren't mathematicians find this hard to believe, but it really is an enormous pleasure when the thing comes out. You think, oh God, isn't that lovely. But the real excitement, which doesn't very often happen, is when these two things come together and you realize that a piece of mathematics actually tells you something about the things you see when you go out bird-watching, and that's marvellous.

'And you see yourself as a theoretician rather than an experimentalist.'
Certainly for the last fifteen years I've been a theoretician, yes.

'And you prefer that to being an experimentalist?'
I think I'm better at it. The world is full of people who are better at doing experiments than I am. I'm a very bad experimentalist. I'm quite good

with animals; I'm a good farmer, which makes me an adequate geneticist because a geneticist has to have his animals healthy. And I can look after plants, I'm a good gardener, and so on. But I'm no good at apparatus.

'So genetics was OK because you don't need apparatus?'

Sure, sure. I was quite a competent geneticist. But the world is full of competent geneticists, and it *isn't* terribly full of people who can do mathematics *and* apply it to animals. The world is also full of mathematicians and many, many people are enormously better at mathematics than I am. What it isn't very full of is people who can apply mathematics to the real world.

'So would you say that you're really very satisfied with what you've achieved? Looking back do you have terrible regrets somewhere?'

Not about my scientific career, I don't think, no. I think I've been extraordinarily fortunate and I've thoroughly enjoyed it.

STEPHEN JAY GOULD *was born in 1941 and is a Professor at*
Harvard University
where he teaches biology, geology, and
history of science.

CHAPTER 11

ROOTS WRIT LARGE

———⟡———

STEPHEN JAY GOULD

Evolutionary biologist

'The male emerges within his mother's shell, copulates with all his sisters, and dies before birth. It may not sound much of a life, but the male *Acarophenax* [a mite] does as much for its evolutionary continuity as Abraham did in fathering children into his tenth decade'.

Stephen Jay Gould: *The Panda's Thumb*

THE PREJUDICE that scientists are neither literate nor willing to explain their work to a wider public is deep seated. The American biologist Stephen Jay Gould is a notable counter-example. His collections of essays, such as *Ever Since Darwin* and *The Panda's Thumb*, combine an enviable style with an ability to make telling connections between widely disparate areas of art and science, and his favourite source of quotation, after the Bible, is Alexander Pope's *An Essay on Man*. Yet he has written of himself 'I am a tradesman, not a polymath. What I know of planets and politics lies at their intersection with biological evolution'. But what intersections! His essays range from a consideration of time's vastness, to women's brains and the Piltdown forgery. Not surprisingly, he does not believe that science is somehow different from other kinds of intellectual endeavour. 'Science is not the heartless pursuit of objective information. It is a creative human activity, its geniuses acting more as artists than as information processors'.

Gould is also noted for his strong sense of social responsibility and, more particularly, for his public opposition to creationism. In the early nineteen eighties, the unacceptable notion that the biblical account of creation is supported by as much—or as little—scientific evidence as the theory of evolution began to gain considerable public credence in certain American states, and the Governor of Arkansas was persuaded to sign a law compelling schools to provide courses in 'creation science' alongside those in 'evolution science'. Inevitably, the legislation was challenged, and Gould was deeply involved in the celebrated trial which followed.

His concern with the relationship between science and politics is a longstanding one. In his book *The Mismeasure of Man,* for example, he presents an extended analysis of the way in which scientific measurement of brain size has in certain instances been unconsciously distorted by the racial and sexual prejudices of the investigator. He once wrote 'I am, somehow, less interested in the weight and convolution of Einstein's brain than in the near certainty that people of equal talent have lived and died in cotton fields and sweatshops'.

A measure of Gould is that he teaches geology, biology, and history of science at Harvard. His main field research is concerned with the evolution of *Cerion*, a group of land snails, but his major theoretical contribution has been the concept of punctuated equilibrium, which he put forward with another palaeontologist, Niles Eldredge, in 1972. They pointed out that on the evidence of the fossil record many species show little change over long periods of time and then disappear, to be replaced quite suddenly by new species. The conventional view is that such discontinuities are merely gaps in the fossil record, and that the evolution of a new species proceeds gradually by the selection of small adaptive changes in form. Gould and Eldredge believe that the discontinuities are real, and that the evolutionary process involves long periods of stasis punctuated by periods of very rapid change. The theory has provoked a great deal of controversy, but has also been influential in determining the kinds of questions paleontologists and evolutionary biologists now ask. The continuous reappraisal of received wisdom is one of the hallmarks of Gould's style. He also questions, for example, how far all the changes in evolution are adaptive, taking many of the cultural aspects of human behaviour as a case in point. These, he argues, arose quite fortuitously as the brain, while evolving to fulfil one set of functions, incidentally acquired a much greater computing capacity than necessary.

The courtroom, and the pages of popular journals are not places where one would necessarily expect to find a distinguished biologist. I wanted to try and learn from Gould something of how and why he manages to occupy such a diverse range of habitats. His book *Ever Since Darwin* has the dedication 'For my father who took me to see the Tyrannosaurus when I was five'. Was this really the event which shaped the rest of his life?

My father was one of that marvellous generation of New Yorkers—the second generation immigrants—who themselves didn't have any opportunity for education, but were very concerned about learning, and wanted their sons to achieve what they couldn't. He went off to war during the Second World War, and when he came back I hadn't seen him for a couple of years—I was five years old—and to make up for it he started taking me places. We went to the Yankee games, that's baseball for anyone who's English, and also to all the museums and other sights of New York. So when I was about five he took me to the Museum of Natural History, and I remember standing in front of that Tyranno-saurus being so utterly scared. A man sneezed while I was there and I was sure the beast was about to devour me. But I ended up at the end of the day deciding to be a palaeontologist. I wanted to be a garbage collector before that because I loved those trucks and the way they whirred around, and the way the garbage got compressed. But I thought it would be great to be a palaentologist, and I never changed my mind. I'm one of the dinosaur nuts who actually stuck with it. It's not rare to be a dinosaur nut, millions of American kids and, I presume, British kids are, but it's rare to stick with it, and I did.

'But what idea had you, at that age, of a palaeontologist or someone who studied dinosaurs?'

Oh, I had no concept of what it was. I thought you spent all your life out in the field collecting bones. I remember at age ten or eleven saying, 'Gee, well, I also want a family and I like cities. Is this profession really going to work for me?', because I thought you had to spend your whole life out in the desert, collecting bones. But at that time my parents were members of a book club and I think because they forgot to send back the card saying we don't want any this month, they got a copy of George Gaylord Simpson's *The Meaning of Evolution*, the cover of which had tiny little pictures of dinosaurs on it. I thought that was great and I tried to read it. I was much too young and I did not understand particularly the last philosophical part of the book, but I was able to grasp for the first time that there was this exciting body of ideas called Evolutionary Theory. Before that it was just the dinosaur bones that intrigued me, just the empirics, these big, fierce, extinct creatures. But I did realize dimly at about age eleven that there was a very fascinating body of ideas behind it all called evolutionary theory and I think that sustained me. If it had just been the bones I would have lost the fascination.

'What was there about evolutionary theory which intrigued you at that age?'

It's hard to say. We all seem to have this fascination with genealogy and roots, and evolution is about roots writ large. It's about, insofar as science can answer those questions, where we came from, what we are, how life evolved, the history of life on earth—I just found it fascinating right from the first.

'Were you a gifted student?'

I don't really know what that means particularly. I did well enough. I wasn't an outstanding student.

'You weren't a prodigy?'

No, no . . .

'I do remember you once telling me a story though, about how you learnt Latin in a few weeks before you went up to university.'

No, I did it by myself in graduate school. It had to do with the Vietnam War. I finished my Ph.D. a few months before my twenty-sixth birthday, and at that time they were drafting oldest first, which means I would have been drafted right away. I already had this job in Harvard and what that means, in short, was that I had to stay in school until I was 26. But I was finished with everything, so I finally said I'd take a Latin course, and so I learned Latin to avoid going to Vietnam.

'But you have learnt other languages too?'

Yes, I like languages a lot. It's history too; linguistics is also genealogical. The humanities have the most interesting analogues with evolutionary theory. It's quite different, because languages anastomose, and join together through cultural contact, which evolutionary branches don't. I don't think there's a good analogue for natural selection in language. But I've found that learning languages has been extremely useful to me. Most Americans don't realize that everything is not written in English and that there are different traditions of thought. For instance, German palaeontology operated under different assumptions than Anglo-Saxon palaeontology, and I've very much benefited, my whole outlook has been altered, by reading material that simply isn't available in English.

'Languages have clearly helped you. But what other skills do you think that you have brought to the study of evolution?'

I have only one strong intellectual skill. I'm not trying to be modest— I'm one of the world's great egotists! I'm not a good deductive thinker at all. I'm not a great observer. I do adequate empirical science. I love working on my land snails—I spend more time on that than anything else—and I think my work is acceptable. But I'm not mathematically inclined. I don't work my way well through deductive arguments, I can never figure out Agatha Christie or Sherlock Holmes stories which are the prototypes of that stereotype. But one thing I'm good at is lateral and tangential thinking. I can see connections among things. That's why the essays I write work. I can see connections that most people think are odd, or peculiar, but are apposite once they're pointed out. And so my work has been integrative, that's what I'm best at doing. I do figure out Dorothy Sayers' mysteries because Peter Wimsey is constructed as that kind of thinker. If you read *Whose Body?*, her first novel, I'm sure that Dorothy Sayers had a theory of thought and that she wrote those novels to counter the Sherlock Holmes tradition that thought was simply deductive and logical. Wimsey just sits down and puts the pieces together at some point, and those mysteries I figure out, because I think the same way.

'I get the impression from reading what you write that you have an enormous store of knowledge with which to do this lateral thinking.'

I think people do have that impression. I'm not badly educated, but people assume, because they see all my literary references for instance, that I must have this immense storehouse of literary knowledge, and it's not true. I think my literary knowledge is adequate. The point is I use everything I know. I do not have a hundred times the information in the depth of my brain. It's just that what little I know I remember, and I see the connections; everything I've ever read I can use. Now that's all. My education is probably quite average in terms of the depth of my knowledge of other fields.

'You say it's quite average, but you also said you remember everything you read.'

I have very good short term memory. When I write an essay I can, for about two weeks, remember everything I've had to read, including page locations, and therefore I don't need to take notes. But I forget everything after two weeks.

'Now, you have said that you think fieldwork is important, and you said earlier that you like to work on your snails. Why do you think it *is* so important?'

Because you lose touch with the material of natural history otherwise. I think Kant was right in that famous old statement that concepts without percepts are empty, and percepts without concepts are blind. If you just ended up working on concepts, you would be indulging in some form of circling around in an ever decreasing radius and eventually would close in on yourself and probably disappear. I think any natural historian has to maintain contact with the empirical world. I would not deny the wisdom of history in that respect. Aristotle dissected squids and Darwin collected barnacles. There's some reason why the great conceptual naturalists have also maintained empirical programmes. Also the snails are beautiful, and when I work on them it's new, it's truly new, even though it may be that only eight or nine people care. For instance, it's been a large mystery for the half dozen people who care about *Cerion* on Cat Island—the snail I work with—which has to do with the fact that there's a peculiar species on the south-east corner. I looked at it, and I said I don't think it's a real species, it looks to me like a hybrid between two forms. The problem is that one of those forms had never been discovered on Cat Island and therefore that hypothesis hadn't come up before. But I'd had enough experience to realize that this was the probable solution. I was able to predict exactly where I'd find this other form that had never been discovered. I went down and it was right there, and it was such a thrill! As I said, not too many people care, it's not going to give rise to any new concepts, but it's clean, it's beautiful, it's right. I've actually found out something about the natural world that nobody had ever seen before. And there's a kind of thrill about new discovery, however small the import of it, that is, at least personally, so satisfying. It may just be indulgence . . . but then there's the other reason of course, that by keeping direct contact with the empirical world your horizons do remain relatively broad. So I would never abandon my empirical research programme.

'But it's curious, because, as you say, you're working in an empirical field where very few people care.'

Few people care about this particular snail. I am working on broader problems. You see, the genus *Cerion* is the land snail of maximal diversity in form throughout the entire world. There's 600 described species of this single genus. In fact, they're not real species, they all interbreed, but the names exist to express a real phenomenon which is this incredible morphological diversity. Some are shaped like golf balls, some are shaped like pencils . . . there's a greater range of form than in any other land snail. Now my main subject is the evolution of form, and the problem of how it is that you can get this diversity amid so little genetic difference, so far as we can tell, is a very interesting one. And if we could solve this we'd learn something general about the evolution of form.

'Just to come back to the idea that the great synthesizers like Darwin need this empirical base to work from. I've always been puzzled why Darwin spent such a long time on barnacles, when it seemed he was involved in so much more important issues'.'

It's hard to say; that's a mystery in Darwinian studies. He spent about eight or ten years working on barnacles in between 1838, when he developed the theory of natural selection, and 1859 when he published it. I think it was largely displacement activity. In his own autobiography he has this wonderful statement about what he learned about the barnacles, and then he has a line at the end, 'Nevertheless I doubt whether the work was worth the consumption of so much time'. There's a lot of reasons why he did it. One is that he had the taxonomist's passion that once you get started on something you really can't stop until you finish it. But the other is that he was afraid of exposing his radical beliefs; not in evolution—evolution's a common enough heresy—but Darwin's own *views* of evolution were extremely radical and there were a lot of reasons why he was quite leery about publishing. He was not, personally, after all a radical man, though his ideas were extremely radical. He had great qualms about publishing something that he knew would be so upsetting and contentious, and I think he largely worked on the barnacles to postpone that day when he would have to face the publication of his views.

'Darwin had a lot of difficulty having his ideas accepted. Have you had difficulty having some of your ideas accepted?'

Oh, I don't think the analogy's really fair.

'I'm not trying to say it's in the same class, but nevertheless you have put forward a controversial theory.'

Yes, and one wouldn't expect that everybody would fall over backwards loving it right away. But I think punctuated equilibrium has done well, especially in setting up controversy, and getting people to redefine the questions they ask.

'Do you enjoy the controversy?'

When it's genuinely and legitimately intellectual, yes. When it descends to pettiness, name-calling, no. Scientists are not immune to that. Science is a funny field, because on the one hand it traffics in ideas, and is commendable in that respect. On the other hand, because the currency of reputation is ideas, which are fluid and hard to define—not like the business world where the currency is currency and it's perfectly clear who's on top and who isn't—you end up in a lot of petty wrangling. There's been more than enough of that about my material and that's most distressing. But good intellectual arguments, of which there have also been many, are always a pleasure.

'What is your image of science, then? How do you perceive the scientific process?'

I don't think it's much different from intellectual exercise in general. Science is distinctive in that its subject matter is the empirical world; scientists must believe there really is an empirical world out there and we can learn about it. So many other professions are not so much working with new knowledge as with interpretation. But beyond that I don't think the methods are outstandingly different.

'But you've argued that Western thought very much influences the nature of Western science. Does that not make the whole thing somewhat relativistic?'

No. I have a very conventional view among historians of science. Radicals in the history of science will actually claim something close to relativism. They may not deny there's an empirical truth out there

somewhere, but it's in the fog, so distantly behind cultural presuppositions that you can never find it, so you might as well not talk about it.
Therefore, for them, the history of the field really is the history of
changing social context and psychological predisposition. I don't take
that position at all. I can't—an empirical scientist cannot. If I didn't
believe that in working with these snails I was really finding out
something about nature, I couldn't keep going. I'd like to be honest
enough to admit that everything I'm doing is filtered through my
psychological presuppositions, my cultural vices, and I think that
honesty is very important because you have to subject yourself to
continuous scrutiny. If you really believe that you're just seeing the facts
of nature in the raw you'll never be aware of the biasing factors in your
own psyche and in your prevailing culture. But that's quite a separate
issue from whether something is true ot not. The truth value of a
statement has to do with the nature of the world, and there I do take the
notion that you can test and you can refute, and so I have a fairly
conventional view about that.

'But you have argued that there is quite a lot of finagling in science?'

There has to be. Science is done by human beings who are after status,
wealth and power, like everybody else. That's why I advocate self
scrutiny. I say, if you don't scrutinize yourself carefully, and you really
think that you are just objectively depicting the world, then you're self-
deluding. The capacity for self-delusion is amazing. To me, conscious
fraud is not very interesting. Oh, it's human interest, it's fascinating, but
the point is, Cyril Burt knew what he was doing. His psychological
make-up may have been skewed in a fascinating way, but he knew what
he was doing when he invented those twins, and therefore, to me, it's
not conceptually very interesting. What's fascinating to me are the
people *do not* realize what they're doing, and that's where you see the
biases of Western thought so clearly portrayed.

'Can you give me an example of that?'

Yes, my favourite example was Samuel George Morton, who's not well
known now, but he was, in the mid-nineteenth century, the major
measurer of skulls. It was he who first established, in a supposedly
rigorous way, the notion that blacks have the smallest crania, Indians
were somewhere in between, and whites had the largest. When you re-

analyse this data, there's no difference. The skulls he measured still survive in Philadelphia, and everybody had about the same size skull when you correct for body size differences. Now, when you actually go through the data you can reconstruct what he did—where he undermeasured, where he made certain assumptions, where he conveniently forgot about body size when blacks had smaller bodies, but remembered it when whites had smaller bodies. I don't think he was aware of what he was doing. He couldn't have been, because it's all in print. If you're fraudulent, you cover up what you're doing. He publishes it all and I assume he was simply not aware of it.

'You think then in some ways science is political: do you think it can be free of these cultural influences?'

No, I don't think it should try to be. As I've said, I think scientists should subject themselves to rigorous self scrutiny. Please remember that cultural biasing factors don't always hold you down, they can be useful. Darwin constructed a theory of natural selection as a conscious analogue to Adam Smith's economics. The principle of individual struggle is the principle of laissez-faire, that if you want an ordered economy you let individuals struggle for profit. Likewise, if you want order in nature you let individuals struggle for reproductive success. Now there's a beautiful example of how cultural context was facilitating rather than constraining. The irony is that Adam Smith's economics doesn't work in economics, but it may be the pathway of nature.

'What about the distinction between science and non-science? You've been very involved in the battle with the creationists. Why do you care so much about that?'

Well, one has to in the American context. You do have to understand that creationism simply isn't an intellectual issue. There is no debate among serious theologians and religious scholars about evolution; everybody accepts it. The Pope accepts it, the leading theologians of America accept it—in fact in the Arkansas Trial which threw out the Arkansas creation law, nine of the fourteen plaintiffs on our side were professional theologians. So it's not an intellectual issue. The reason why we have to fight it is very clear. It's a direct attack upon my profession. If these guys win, evolution, which is the most exciting concept in biology, and is the integrative concept of all the biological sciences, will not be taught. What's more, creationism is an attack upon all science, not just my narrow little field. These guys believe in the literal word of Genesis, and if that's true, everything goes. If the Earth really is 10 000 years old then all astronomy is wrong, and all of cosmology is wrong, because the astronomers tell us that most of the stars are so far away that their light takes longer than 10 000 years to reach us, therefore there's something wrong if the universe is only 10 000 years old. All of physics and atomic theory goes, because if radioactive dating consistently gives these ancient ages for old rocks then, as it's based on the fundamental behaviour of atoms, there's something wrong with that knowledge of atomic structure, if it's all a delusion and the Earth is only 10 000 years old. So it's an attack on all of science; it's an attack on all of knowledge, because these guys would substitute an authoritarian source, namely the Bible, for any self enquiry and investigation. That's a very broad attack on anything intellectual. It's not just debate about one part of biology. So we have to fight it.

'What do you feel about other fringe science, like extra-sensory perception and so on?'

My general feeling is that it's a source of endless frustration to me. I'm absolutely convinced that 98 per cent of it is the work of cranks and kooks, and people who just aren't very rigorous. They may be nice folks and even have good ideas. The problem is that it is not inconceivable that there's a couple of per cent of that stuff that's good and could be exciting and could really be fundamental, but life's short and I'm not a magician. The only people who can really test these claims are magicians who know what all the tricks are. I cannot get into that field. It's something utterly beyond me. There's nobody who can be more easily tricked than a pompous scientist who thinks he can spot anything. That's why Uri Geller couldn't fool a carnival magician, but he fooled a lot of scientists.

'But I sometimes get the impression that people want to believe in these ideas.'

Sure they do.

'Why do you think that is?'

Oh, I think it has to do fundamentally with immortality doesn't it?

Freud was right. So much of human life is our attempts to deal with this most terrible fact—which our large brains allowed us to learn for non-adaptive reasons—namely that we must some day die and disappear and not be here any more. There's such a great desire to believe that mind is transcendent of the universe in the hope that, although we know our bodies disappear, there's something about us that will be immortal. If there is ESP, if mind can be transferred, if mind can live, if there is reincarnation, then we might have continuity. That's what it's all about fundamentally.

'You are known as a writer and a teacher. Does it matter very much to you, the teaching and the writing? Why do you do it?'

I don't mean this to be mis-taken . . . I really do not have a burning passion to educate the public, although I think that's very important. I mean, insofar as the public is educated through the lectures and essays I write, I think that's a marvellous side consequence and I'm certainly a great believer in it. And when I write, I try and write in a way that will be accessible to everybody. But basically I write those essays for myself. That's the only reason why they can have any freshness, and decency. I write them because the world is so endlessly fascinating. There's so much to learn, and every time I write an essay, I learn something. That's why I do it. I do it for me.

'Do you really mean to say that you write them for yourself? Do you do your thinking through writing?'

To some extent. The point is that, given the norms of the scholarly enterprise, I cannot write papers on all the subjects I treat in my popular essays. For instance, last month I wrote an essay on Philip Henry Gosse, the author of *Omphalos,* that wonderful 1857 work claiming that God created the earth with all the strata having the appearance of pre-existence. Gosse was a fascinating man, but I'm not a Victorian scholar. Where would I come off as a palaeontologist writing a serious essay with 150 footnotes; who would publish it? Therefore, if I didn't have this format of the popular essays, where would I ever do it? I wouldn't have the impetus.

'So the impetus is really self-education?

Sure. Sure. I'm perfectly selfish about those essays. I write them for myself.

'Now, how do you actually write them, because I've seen it said that you write one draft only.'

Oh yes, it has to be done that way. If I couldn't do them that way I'd have to stop, because I have this whole other career with my snails and with evolutionary theory. Fortunately I can write quite quickly. As I said, I do have only one intellectual skill, but I don't mean to degrade it in any way, or be falsely modest—I'm not a modest person—and that is I do see connections. But through that ability to see connections, I can be very clear about outline. I can, as I said before, remember everything for two weeks and then I forget it all. I'll do the reading I need to do and then—and I say this somewhat metaphorically, although I truly do believe it—I'm a Platonist with respect to outline. I think I do believe there is a correct outline for every piece, one proper outline. It's up there in heaven and it's just a question of finding it. And so I do all the reading, and then I sit down and I wait for the outline to come to me. It always does, and when it comes I write down the outline quite extensively. Now, once I have the right outline, that's the way the essay should be written, and then I can just write it you see. So I can write it fairly quickly; I'll do it in an evening. I don't want to be absurd—I'm not like Mozart, I don't simply sit down and write out the essay, with no mistakes. No, as I write, there's a lot of crossing out, every sentence I write has crossings out and rewrites. But once the sentence is down, it's finished. I don't switch paragraphs around, I don't write a second draft. Once it's done it's finished and it's ready to go out.

'Now what is the relationship between art and science in your life? You like the arts?'

Yes. I'm hopeless at the visual arts. I'm an appreciator, a visual appreciator. I'm one of those people that really can't draw a straight line on a piece of paper. My own personal involvement is in music—I'm a singer. I play some instruments, but not well. My immediate activity is singing and, indeed, had I had a better voice that would have been my dream. As I keep saying, my two unfulfilled dreams are to play centre field for the Yankees, and to sing Wotan at the Met. But not being endowed with skills in either of those directions, I've had to do other things!

'How do you perceive opera in relation to your other interests?'

I suppose what I like about it is it combines everything—it's singing, it's music, it's drama, it's vision, it's dance. I guess I'm an integrationist at heart.

'I can see that, but what about baseball?'

I grew up in New York—that's personal history! My grandfather assimilated to America on baseball. I'm a third generation Yankee fan. I grew up in New York City at a time when there were three great baseball teams, two of whom would win the pennant and play in the World Series almost every year. My childhood hero was Joe Di Maggio, the most graceful man ever to play baseball. He's still a beautiful man to this day doing advertisements for Mr. Coffee and Bowery Savings Bank. He still has that elegance and grace that was so evident on the ball field. No, I was just a New York City street kid and we're all baseball fans. That's personal history.

THE SEARCH

ANTHONY EPSTEIN *was born in 1921 and is an Emeritus Professor at the University of Bristol. He is now in the Nuffield Department of Clinical Medicine at the University of Oxford, and is a Fellow of Wolfson College.*

CHAPTER 12

BETWEEN THE LINES

——⊸∘◉∘⊷——

ANTHONY EPSTEIN
Virologist

SIR PETER MEDAWAR has pointed out that the scientific paper is a fraud. It gives a completely misleading impression of the way scientists work. The formal presentation of orderly progress gives no idea of how much of the process is disorganized, or of the failures and wrong turnings along the way. The popular image of how discoveries are made, particularly in biology, comes from a few, probably atypical, cases, such as Fleming's apparently serendipitous discovery of penicillin, and Watson and Crick's discovery of the structure of DNA. How does it really happen?

Anthony Epstein and his research assistant, Yvonne Barr, discovered what is now called the Epstein–Barr virus, the first virus to be shown to cause cancer in humans. Success came after more than two years of failure, and it took even longer for the work to be widely accepted. On the basis of Popper's view that what counts in science is not the repeated demonstration that something does exist but the failure to prove that it doesn't, they should have given up. So what kept them going? How far was the eventual discovery the result of calm reflection and grand experiment? Did fortune play into their hands at all? How much was inspired guesswork, and how much grim determination?

Viruses are not really living things. They are minute infectious agents, sequences of genetic material, which remain inert until they enter a cell. Then they hijack the cell's genetic machinery and force it to manufacture more virus. The actual effect on the cell depends on the circumstances and the type of virus, but in certain instances, the integration of a virus into the cell's DNA causes it to become cancerous.

The idea that viruses play at least a part in causing some types of cancer is now generally accepted. But until very recently this was far from the case. The notion that cancer might be caused by a virus goes back to the beginning of the century, when an American physician, Francis Peyton

Rous, discovered that a particular tumour in chickens seemed to be caused by a virus. So unlikely and controversial an idea did this seem in 1911, however, that he did not dare call the agent a virus. It was another quarter of a century before it became known as the Rous sarcoma virus, and even then the idea that viruses might have an important role to play in causing malignant disease was still not widespread. When Epstein first began work, just after the last war, very few people believed in, or were even interested in the possibility that they might be involved in human cancers.

The cancer that Epstein linked with a virus was Burkitt's lymphoma, named after Denis Burkitt, the British surgeon who first identified it. Burkitt worked out in East Africa and in the late 1950s realized that the many seemingly different cancers of the lymphoid system he was seeing in African children were in fact the same tumour, and that the distribution of the tumour was related to temperature and rainfall. This strongly suggested that some infective agent was involved, and Epstein finally succeeded in showing that this agent was a virus belonging to the herpes family, whose other members cause cold sores, chicken pox, and shingles.

Rous had demonstrated that his virus caused cancer in chickens using the classical techniques of virology. He took tumour tissue, ground it up, and extracted a fraction which he then injected into disease free chickens. These chickens then developed tumours. In other words, he demonstrated the action of the virus by producing a biological effect. Epstein's approach was rather different. For the ten years or so since the war he had been exploring the possibility of using a new technique to study tumour viruses. This was electron microscopy, which had only just become available. Unlike a conventional microscope which examines objects by means of light, the electron microscope uses a beam of electrons, making it possible to observe much smaller objects than with the light microscope. Viruses had previously been too small to observe directly, but the idea that they might now be shown to be present by an essentially morphological technique, rather than by their effects in the body, or on cells in tissue culture, was regarded as heretical by most virologists.

––––––

At that stage I was doing something which people considered really rather bizarre and insane. I was working on the Rous sarcoma virus, which induces tumours in chickens, and that was at the time a very unfashionable thing to be doing. There were probably about half a dozen people in the whole world who were interested in chicken tumour viruses.

'Why did you then choose that?'

After I'd qualified in medicine, I went into the Army and there was a patch during my military service, after the war, when I was in charge of wives and families on a station in North India. During that time I became rather disenchanted with the sort of thing one was expected to do in general practice, and when I came out of the Army I decided I would rather be in laboratory-based medicine than in any branch of clinical medicine that I'd come across.

'And did you start working on cancer straight away?'

No, I did a very short and limited training in hospital pathology at The Middlesex Hospital Medical School, in the Bland Sutton Institute. But they had a tradition of this kind of work from before the war, and at the time I came out of the army they were just starting up again. So after this very limited training period the opportunity arose to move over on to the research side and I grabbed at it instantly.

'So it was chance that you started research?'

Yes, it was just chance.

'And were you doing classical virology at that stage?'

Yes, but one was applying new techniques, new ideas. We had a combination of ways of looking at how much virus you had in your preparation and I suppose the innovation was to compare looking for biological activity with what you could see in the electron microscope. I think it was really the first time that this had been done at all. I was using the electron microscope in the very, very earliest days of its application to biological material, and that was quite exciting.

'How did you come to be using this new technique, the electron microscope?'

There was one of the very earliest electron microscopes in the Institute, extremely crude and difficult to use, but it was there. And I started to use it.

'So it wasn't that you thought you must go out and get an electron microscope?'

No, but as soon as I became aware of what it would do, then certainly there was this extremely strong feeling that this was something one must find out about. And it became quite obvious that one had to go to the fountainhead, the place in the world at the time that really was the Mecca for this kind of work, albeit totally unrecognized by the great mass of biological scientists. This was the combination of Keith Porter and George Palade—who now has a Nobel Prize—at The Rockefeller Institute, as it then was, in New York. Their pioneer studies were just coming to fruition in the middle fifties, and I went in 1956.

'Was that visit very important to you?'

Very, very important, yes. And enormously exciting. There one was, involved right at the centre of something which was clearly of fundamental importance to major aspects of progress in biology in many different fields. And it was just a couple of dozen people in the whole world who seemed to know this, though one was absolutely sure oneself, and it was really great . . . very exciting.

'So you came back very excited. What happened next?'

Well, I naturally continued to apply these various techniques of electron microscopy coupled with biological activity, and we introduced some new things. Then in 1961, to come back to this question of chance which you were talking about a minute ago, somebody called Denis Burkitt came and gave a talk at the Middlesex Hospital Medical School. Of course, in 1961, nobody had the faintest idea who he was. He was, in fact, a surgeon from Uganda—he described himself as a 'bush surgeon' actually—and because he had a connection with people in the Department of Surgery in the Medical School he used to give a talk there every three or four years when he came on leave.

'So it wasn't even in your Department that he gave the talk?'

Oh, absolutely not. He came and talked to students and young members of staff and usually, like most speakers from developing countries, he would show exaggerated examples of this or that kind of disease, you know, the largest hydrocele in the world and so on—all very dramatic. But this time he was talking about the commonest children's cancer in tropical Africa. I don't know why, but I caught sight of the notice of his talk and I just went and listened to him. That, of course, was again an extremely exciting high point because he talked about this new tumour, which, of course, is famous now and carries his name—Burkitt's lymphoma. It was the first description of the tumour ever given outside Africa, and in addition to telling us about the tumour, which was very bizarre and peculiar and made us think in terms, perhaps, of a virus induced tumour, he also told us of his second major contribution. Most people only make one, but he made two right off at the beginning! And his second major contribution was to do the epidemiology of the tumour and show that its causation was dependent on temperature and rainfall. Now, anything which has geographical factors such as climate affecting its distribution must have some kind of biological cause. I'd been listening to this for 20 minutes when I was hopping up and down in my seat, absolutely certain that this had to be a human tumour which must be investigated for a virus causation.

'Was there evidence for any human tumours being caused by viruses at this stage?'

Oh no. There was little evidence, even, for any animal tumours being due to viruses. It was just great good fortune that I had worked in this field, because most people would have regarded the whole idea as ridiculous. So I decided instantly that it had to be investigated. We had Denis Burkitt round to tea and made arrangements with him to see if I could get biopsy samples from his patients in East Africa sent to us here at the Middlesex Hospital Medical School.

'So from the moment you heard that seminar you decided that you were going to work on this tumour?'

Yes, absolutely. From that very moment.

'And you dropped everything else?'

Everything else, yes. It was very exciting.

'And were there problems in switching? What about money?'

There were no problems with switching in the lab. Neither were there problems in getting support. Indeed, the British Empire Cancer Campaign, as it was then called—it's now the Cancer Research Campaign—became interested. There was, however, a great deal of difficulty in scientific politics, in getting a foot in the door to get material from these tumours from Uganda.

'I don't understand the problem.'

Well, there was obstruction. There were people who wanted to corner this material.

'You mean, other people recognized its potential?'

Yes, other people had become interested in this tumour. There was a great deal of high level difficulty and unpleasantness, which was finally resolved by the Cancer Research Campaign very generously paying for me to go out on a reconnaissance trip to see how we could set things up to get this material.

'Were you then the only person having access to this material?'

No, there were other organizations who actually set up labs in Uganda in order to investigate it. But our material was sent to us by air, and I suppose three-quarters of the shipments reached us safely. Every now and again, particularly in winter, there would be fog and planes would end up in Manchester instead of Heathrow, and we wouldn't get the stuff until three or four days later, and it would all be useless. But we slogged away at this as best we could.

'What do you mean by "slogging away at it"?'

Trying to investigate the material for the presence of some quite unusual virus. We started first of all with all the standard isolation procedures and came up with absolutely nothing. We looked at the material in the electron microscope, but again there was absolutely nothing.

'Weren't you a bit demoralized by that?'

Oh, it was frightful. I mean, having uniformly negative results for two or three years at that stage in one's career was an extremely alarming situation.

'Did you have a permanent job at that stage?'

Oh no, no. Very few people had permanent jobs, one was hanging by a hair from year to year. It was a very harassing time.

'So it's to the credit of the Cancer Research Campaign that they continued to support you although your results, your annual reports, presumably said "sorry, nothing to report".'

Nothing . . . zero . . . nil to report, yes. It was very frightening.

'And what did you do each day?'

You came into the lab and waited for the specimens to arrive. You hoped with bated breath that they wouldn't be delayed, that this shipment wouldn't be contaminated. Then you tried over and over again to grow the thing in tissue culture. After all, you didn't just want a specimen, you wanted this stuff to be growing in order to have it in the lab for continuous investigation.

'Did it grow for a short while and then die?'

Well, it really didn't grow. I mean, it was a mad thing to try and do because people had tried to grow members of the lymphocytic series of cells without success from the very early days of tissue culture at the beginning of the century. But, you know, we just battered on at it.

'And did you try and change the recipe for the growth medium?'

Oh yes. We changed the technique, the growth medium, guessing all the time. We tried growing them in clots of plasma, growing them as little lumps on grids in fluid medium, even growing them on tea-bag paper, I seem to remember, floating as a raft, in the hope that the cells would perhaps drop off and fall through and grow, and so on.

'But were you doing other things like trying to extract a virus that would infect mice.'

Yes, by inoculating mouse brains, inoculating developing eggs, all the standard things.

'You were doing different things each one being a more miserable failure than the last.'

Total failure, yes. Total failure.

'For how long do you think you would have gone on?'

Well, I don't know. You had to be pretty dogged. But it had to be right. It just had the feel of being right. And that's why one carried on.

'So when did you actually begin to make a little progress?'

Well, I can remember exactly how it happened. It was one of those shipments, which for some reason, either to do with weather or delay on the plane, was late. We didn't get it first thing in the morning as we usually did, but very late on a Friday afternoon and, of course, everybody was waiting to go off for the weekend. I remember distinctly looking at it, and Yvonne Barr, who was working with me at the time, said 'Oh, look, it's contaminated, let's throw it away', because the fluid in which the material travelled was all cloudy, as if bacteria had grown. Of course, many of these tumours came out of people's mouths and jaws and, therefore, not surprisingly, were contaminated with the normal bacteria from the mouth, which made things unworkable. I don't know why, but I just didn't want to throw it away. So I had a quick look at this cloudy fluid under the microscope and to my amazement, it was cloudy, not because of bacteria, but because of huge numbers of single cells which were floating there in the fluid. They had been shaken out of the piece of tumour. And immediately it reminded me of something I'd seen years before when I'd been visiting Yale University School of Medicine, where they had been growing tumours in a rather special way. I suddenly realized that maybe we were doing all this wrong by trying to grow little lumps of tumour as was the standard thing in those days. Maybe we should be breaking it up into these single cells which was what the people at Yale had done with their tumours. So, we broke this stuff up into a suspension of single cells and grew that in tissue culture and it worked.

'You hadn't been able to grow the tumour until that stage?'

No, never before. That was the first time any member of the human lymphocytic series of cells had ever been grown in continuous culture, and that was the major breakthrough.

'But that, of course, didn't really solve the virus problem?'

No, that didn't solve the virus problem because, naturally, we slogged our guts out again to demonstrate virus in these cultures by all the standard biological isolation techniques and the results were totally negative. But we were very lucky once more in that I had become familiar with, and appreciated the use of, electron microscopy in combination with biological work, so we were never satisfied with a negative biological test. I always wanted to look and see. As soon as we had enough of this extraordinarily precious cultured material to spare, we embedded it, sectioned it and looked at it in the electron microscope.

'And . . . ?'

And there, in the very first grid square was the cell with unequivocal virus particles. There was no doubt whatsoever. It didn't tell us what kind of virus it was. But we knew the family by its morphology. It could have been some quite ordinary member of that family, but, on the other hand, biologically it was not registering any of the features of ordinary members of that family. Therefore it was extraordinarily exciting.

'What did you actually feel when you saw it?'

I switched the microscope off in case the specimen burned up and I went out and walked round the block two or three times before I dared come back!

'What did you think as you walked round the block?'

Well, I knew that this was it. It just had to be right.

'And, of course, it *was* right.'

It was right, yes.

'Now did people accept this quite readily?'

Oh no, no! It was a tremendous uphill battle from the very beginning. First of all, conventional virologists in those days would not accept that something seen in the electron microscope was a virus. This was an absurd idea as far as I was concerned because for decades people had been looking in the light microscope and saying, oh, yes, this is a bacteria and we can tell what family it belongs to. But in those very early days of electron microscopy, the idea of using it to identify a virus was unheard of, particularly in the case of an agent which had apparently no biological activity. It couldn't be a virus, it must be an artefact, or what people tended to call 'virus-like particles' which shows the degree of sceptism they had at that time.

'It must have been quite demoralizing, now that you finally had the virus, or at least morphological evidence of the virus, and still people said "Ah, but this is not the real thing".'

Well, I knew it was right, so I really didn't care. But—and it's a peculiar thing—it's been a very strange story with this particular virus. As one of my French colleagues who has worked in this field for some years said, the EB virus—the Epstein–Barr virus—is 'mal aimé'. It's unloved, it's had difficulties from the very beginning. I think all that has really been dissipated in the last few years, but it was a tough struggle at the beginning.

'And how did you cope with that struggle?'

Well, I remember it was a strange business. We needed some biological work doing because we didn't really feel confident about the way we'd done it. I mean, we'd done it all right, but the fact that it was all negative meant one felt, well, it should be done in some other laboratory. And I remember going to two leading workers in the herpes virus field, because EB virus is a herpes virus, in this country and saying 'Would you collaborate with us? Would you look at this material and see if we're missing something, or have made some ridiculous mistake in the lab? Is it some quite ordinary herpes virus?'. Neither of them wanted to know. I remember Yvonne Barr was very upset that in fact we had to send it abroad, to America, to collaborate with colleagues at the Children's Hospital in Philadelphia, in order to pursue these further studies and get some independent confirmation of our negative findings.

'It's a remarkable story. If you think about Popper's concept of science, he says scientists try and *falsify* their theories, rather than trying to prove them. You spent two and a half years repeatedly falsifying your hypothesis, because by all the criteria of classical virology, you did not have a virus.'

Exactly. That was, I think, the source of all this difficulty at the beginning. But of course, as far as we were concerned in the lab, the fact that it wasn't behaving as such a virus should have behaved was the best result we could have got. It became absolutely clear that this was something completely biologically new and distinct.

'You said earlier that there were other people who were setting up laboratories in Africa. Why was it you who managed to solve the problem?'

Well, I think the answer to your question is absolutely straightforward. People applied the techniques of classical virology to this system, and this was a virus for which there was no biological test. We know now that there *is* biological activity, but it's of quite a special kind and not anything that anybody was looking for then. If you applied classical virology to such a system in those days you came up with nothing. It was because of the electron microscopy allied to biology that we found it, I'm absolutely sure. And that in turn, was due to the excitement and stimulus of having been with that small band of people at The Rockefeller at the crucial time, in 1956, so that when 1963/4 came, one knew exactly what to do and most people weren't into that at that stage.

'You give yourself rather little credit other than persistence. It's almost as if it was chance that Burkitt came, and that you went to Palade at the right time.'

No, it wasn't luck that I went to Palade. I went to Palade on purpose at the right time. It was luck that Burkitt came but I think it was serendipity, not just chance, that made me follow it up. It was a conscious decision after hearing him speak, but it was certainly luck that I heard him.

'Well, it is as Pasteur said: fortune favours the prepared mind. What do you think your skill is as a scientist? You're not a theoretician?'

No, not at all. I don't understand any of that. I think just sort of messing about is the answer. You've got to keep messing about at the bench. You see how to change this just a little bit, and you see how to change that a bit, and you want to tinker with something and find a slightly different and new way of doing it. You make a little bit of apparatus . . . .

'You actually like using your hands?'

Yes. I think that's frightfully important. You've actually got to be there, seeing it, messing with it . . . .

'Getting a "feel" of the material.'

Yes. It's very difficult, but I think it's very important. I'm sure it's not right, say, in astrophysics or big science, but it's very important in certain kinds of biological work. It means registering inside yourself minute changes, tiny things which may have a big influence.

'Like you remembering, when you saw those separated cells in the specimen from Africa, that you'd seen something like it before.'

I'd seen it in a different circumstance, completely different. But the fact that I looked at the thing and saw it with my own eyes as a suspension of separated cells, meant that I immediately made the flashback to the separated cells that they had grown at Yale.

'Do you still do things with your own hands as much?'

Not really, no. Well, I suppose I do do some things but not very much, and nothing like as much as I would like.

'But you direct. Is that less satisfying?'

I think perhaps it is. But on the other hand they're good people and things go well and one has a feeling also, though it sounds absurd, that there is a tradition that has to be passed on—a way of doing things. There are different ways of doing things and I think that certain kinds of meticulousness and fussing over details are important and should be passed on to younger people.

'Who did you learn your concern with minutiae from?'

I don't know. I suppose it probably comes from an obsessional sort of temperament. Nobody ever really taught me anything until I went to George Palade.

'Was he obsessional in this way?'

I suppose you could say he was, yes. But I've always been like that.

'It must be enormously gratifying to discover a virus of such enormous importance. Do you think the boosting of one's ego is an important feature in science?'

Well, I suppose success in any human activity gives satisfaction. I mean, it's certainly better than failure, to put it at its lowest.

'But do you think it's a prime motivating factor?'

It's awfully difficult to answer that question. One often wonders that about people who are pre-eminent as, say, musicians or in art and so on. There's obviously something there which makes them do whatever it is they do, but I'm quite sure that there is a drive to succeed as well, and it's fuelled by the satisfaction of doing so. It has to be, we're all human.

'But say you could have discovered the virus but had to remain anonymous?'

It wouldn't have mattered. Just to get the damn thing. That would have been great!

WALTER BODMER *was born in 1936 and is Director of the Imperial Cancer Research Fund, London.*

CHAPTER 13

SHOOTING AT THE MOON

⎯⎯◦⊙◦⎯⎯

WALTER BODMER
Geneticist

SCIENTISTS may or may not be a solitary breed. Either way, the days of lonely researchers in scientific garrets are over. Not only can they no longer afford the equipment but, in many cases, a concerted effort simply makes more sense. This is certainly true in a field such as cancer research, where understanding the origins of the disease demands a variety of approaches, and where there are also the allied questions of prevention and treatment.

One of the most prestigious and successful institutions devoted entirely to these problems is the Imperial Cancer Research Fund in London. Here, under the auspices of a single organization, many laboratories conduct research into widely differing aspects of the disease. Directing research of this scale and diversity is a very different activity from pursuing an individual research programme, and even from running a single large laboratory. I wanted to find out what doing science in this way involves, and what kind of scientist takes on such a job.

The ICRF's Director is Sir Walter Bodmer, who came to the post from the Chair of Genetics at Oxford. That someone not medically trained, whose work was not primarily concerned with cancer, should be running a cancer research institute seems, on the face of it, surprising. Yet cancer is essentially a genetic disease, arising, at least in part, from a defect in the genetic machinery which controls the life of a cell. Unravelling what factors may be involved in producing or triggering such a defect, what actually happens in the cell to cause it to proliferate in a runaway fashion, and what the therapeutic possibilities might be demands a range of scientific and medical expertise ranging from cell biology and genetics to immunology and statistics.

Bodmer trained originally in mathematics and switched to genetics as a postgraduate working under R. A. Fisher, the pioneer of the application of statistics to genetics and a major contributor to the field of population genetics—the study of how genes are distributed in a population. To his skills in mathematical genetics he adds those of molecular biology. He worked with the American geneticist Joshua Lederberg, who was the first to show that bacteria were capable of sexual reproduction, and so transformed the study of bacterial genetics. Much of Bodmer's recent research has been concerned with the body's immune system and, in particular, with the genes which enable the body to distinguish between 'self' and 'non-self', such as a tissue graft, or between a normal cell and one which has been changed in some way, say by infection with a virus. It is armed with these disciplines that he has taken on a problem which defied the blunderbus assault of President Nixon's much trumpeted campaign to find a cure within a decade. Unlike aiming to put a man on the moon, understanding the problem of cancer requires not so much technological development as fundamentally new insights into the way cells behave.

Fortunately, however, it is still possible to design therapeutic measures even though the causes of cancer are not yet fully understood. It was the dream of Paul Ehrlich, one of the founders of immunology, to find a 'magic bullet' which would seek out and destroy diseased cells or parasites without damaging healthy tissue. One of the most promising avenues of research now is the use of immunological techniques to try and do just this. If the surface of a cancer cell is even slightly different to that of a normal cell then, in principle, it should be possible to target and destroy that cell leaving the normal cell intact. The recent development of monoclonal antibodies offers such an opportunity. Monoclonal antibodies are agents with a high degree of specificity, which preferentially bind to the surface of particular cells. If it proves possible to use antibodies to deliver a lethal drug specifically to cancer cells, then the magic bullet will have become a reality.

Bodmer has always been interested in the wider issues of science, particularly its social aspects, and this has not changed with his directorship of the ICRF. In the past he has been concerned with the validity of ideas about the genetic basis of intelligence as measured by IQ tests, and, very recently, he chaired the influential inquiry sponsored by the Royal Society into the public understanding of science. Clearly he has not found the job confining in this respect, but in many other ways it must, surely, represent a considerable loss of scientific liberty. Why, I wondered, had he been tempted to leave Oxford and the freedom of pure research to take on such a formidable range of applied scientific and administrative problems?

————————

I thought it was an interesting challenge and an opportunity that I simply couldn't pass by. I enjoyed Oxford immensely and it gave me opportunities in my research that I valued greatly, even though I'd come from Stanford before that. But at the ICRF you've got more scope. It's a large, broadly-based research institute, with a large variety of subjects being worked on, and both in terms of my own research, and in terms of the broader challenge of helping a field like this along, I felt it was something that I just had to take on.

'But are you still active in research at ICRF?'

I am, and I think that's extremely important. One might ask 'In what way are you active?' I don't actually hold test tubes very often—the people in my lab will testify to that—but I have a lab which is immediately adjacent to my office. I have people with whom I work closely, and whom I see more or less on a day to day basis. I think that it's essential to maintain an active research interest for a variety of reasons. One of the first is for oneself, selfishly. I think you've got to maintain your own research interests and liveliness in that way. I think it's very good to have a research lab which can to some extent be an example to others, and which can form a base sometimes for starting new things—you can get something going in a small way, and then transfer it. It forms a basis for collaboration with others. I think it's right in all sorts of ways. Not everybody agrees with this, but I feel that in a position such as mine you've got to remain in some way an active scientist in order to carry your own self-respect in the field, and that of others.

'It's almost a standard against which others can measure themselves?'

In part, yes. I would find it difficult I think, to deal with many of the things I'd like to deal with, if I didn't still have that as a nucleus of interest.

'But as director of the ICRF, what does it mean to direct a large institution towards understanding cancer?'

It means so many things. A lot of it is dealing with people and their problems—often personal, but also, of course, scientific. It's thinking about areas of research and trying to promote them, and encouraging people in particular areas. It's also appointing people, of course. But a lot of it is dealing with people and scientific problems and their interactions and relationships.

'To what extent do you choose the problems for others? You spoke about how you could try out new ideas in the Director's unit as it were, but to what extent do you determine the overall direction in research?'

Well, that's difficult to say. Obviously we've got—and should have—many distinguished senior scientists in an institution of this kind, and the whole aim of choosing them is that they are people who have good ideas and do their own research in the general context of a cancer research institute. So you don't sit there and tell people what to do at that sort of level. But, of course, you have a special role when it comes to choosing new areas, if you've got a new appointment to make. Also, when it comes to encouraging people. A lot of it is getting their confidence, talking to them, maybe pointing them in certain directions, maybe bringing people together, seeing what's needed to make things work more smoothly, and so on.

'How do you know which direction to go in then?'

You *have* to keep up with what's going on in the field—and if I've done anything in the last few years, it's been to learn a great deal about cancer—because you must become, and have to be seen as, an expert in the field, which I can't claim to have been before. Obviously, many of the areas of work are closely related to my former interests, but you really have to become widely knowledgeable–that's important. You have to be prepared to take advice and consult with many other people. Of course, I don't make the decisions individually myself; I'm responsible to the Council of the ICRF which includes distinguished scientists. So it goes through a process of thinking about problems oneself, talking to other people about them and finally putting proposals forward.

'Is the process different to when you were working in Oxford? Are there more constraints because of doing, as it were, applied research, or do you not see the Fund as doing applied research?'

Well, I think the goals of a cancer research institute are very, very important. It's very easy to come from my sort of background and say 'Oh well, just go on doing what you were doing before' and call that cancer research. I don't believe that one can do that at all. We depend on the public for our contributions, and we have a great responsibility because of that. So we must look towards the applied areas. At the same time it still is true that much basic research is needed. So you need an appropriate balance. I would, though, be as worried about missing an opportunity for an application as I would be about not doing enough basic research, and that does make us different from a university.

'Is that a hard pressure, do you think?'

No, I don't myself. I think applied problems have just as much interest as basic problems. And I think very often in the history of science the two have stimulated each other, so I don't worry at all that we actually have a goal and a direction. I think it's fortunate that in the course of pursuing that, we can also do a lot of interesting science.

'Do you think administrative skills go along with scientific skills?'

Well, in general there's no reason to believe that they should, and I should explain that at the ICRF the fund-raising side, and certain aspects of the administration, are dealt with by the Secretary of the Fund, who is my parallel on the non-scientific side. But when it comes to dealing with the science—scientific administration—this must depend on a good knowledge of science and its evaluation. And while perhaps not every scientist is an administrator, I think the person whom you get to direct an institution like this must be both a scientist and capable of dealing with people. I think if you don't have the knowledge and the skills, and intuition about the science, I don't see how you can really manage something like that properly.

'But where do the administrative skills come from?'

I think these are things that you acquire gradually as you progress through your career. And of course, an academic career is a changing thing. You start by working all the time at the bench or doing your own problems. You gradually acquire more responsibilities in one way or another, you're on committees of one sort or another, you gradually obtain more administrative experience, especially in a university setting. I think that's what brings it to you, and some people manage, or seem to manage, better than others. And, I think, managing a group, a team, even within a smaller laboratory is an administrative problem. It's on a larger scale in a larger institute, but you still get that experience working with a smaller group.

'To what extent is science nowadays about big groups? I mean, physics has become big science. Is biology, cancer research, in that class?'

I don't think so. But I think one has to consider very carefully whether there aren't directions, say, in cancer research, where, if you put in the sort of effort that was put into putting a man on the moon, you would at least solve some of the problems. I don't think we should be too complacent in assuming that that may not be the case in the future. But nevertheless, it hasn't been so far. In fact, in our institution, the basic unit is generally not a very large one. Each member of the permanent scientific staff has their own laboratory and a fair amount of independence, and I believe the way to create the larger units that to some extent are needed in an area, is by bringing together laboratories with related interests. It is very difficult nowadays for a single small laboratory to make headway in a major competitive field.

'Why isn't cancer like putting a man on the moon? There was, after all, Nixon's "Let's solve cancer, like putting a man on the moon".'

Well he did something which I think one should never do, and said 'We'll solve the problem in ten years'. That's a definite period of time, and I think one should never do that—you'll always make a mistake if you do. Which isn't to say that we won't solve the problem. I'm convinced that at some point in the future cancer will be a problem that we can deal with like we now deal with infectious diseases. But if you ask me 'When?' I would be a fool to put any sort of time-scale on that. Now, you're asking why isn't it like putting a man on the moon? I once said 'Well, the thing is we know there's a moon, but we don't know in what direction to go to get there. And somebody said to me, 'Well do you actually know whether there's a moon?'! I think we're getting a little closer to the situation where we know there's a moon, and there are certain directions that we could go to find it. And, I think, as I said earlier, one has to be a little cautious in assuming that we're so far away from the possibility of directed solutions that we really can't do anything about it.

'What is the moon for cancer research? If the good fairy came along to you and said "Professor Bodmer, I'll answer any three questions for you", is one in a position in cancer research to know what those three questions are that would really give you the answer?'

It's always a two-sided story. The moon is made of two sorts of cheese—there's prevention, and there's cure and treatment. And they're often put in opposition to each other, which I think is a great mistake, because they're complementary and you need research in both; the future is certainly going to lie in dealing with both. So, the more you can prevent the better. It's often said that up to 80 per cent or more cancers are in some sense environmentally determined, but it's then assumed by some that that means you could relatively easily deal with 80 per cent by controlling the environment. I think even those people who have done the original research, very distinguished epidemiologists such as Sir Richard Doll, would say that you will only ever be able to deal with a part of it. So whatever else you'll do, you'll still have to deal with a cure. So, in each case there's going to be a variety of problems.

On the prevention side, one thing I would love to do is to stop people smoking in some way. You can't do it legally, but I would like to see much greater emphasis put on a general antagonism to smoking. Things that are still done by tobacco companies, like sponsoring sport, for instance, gives them an advertising potential which perhaps is unwarranted. There will be other areas of prevention which one can't yet predict. There might be manipulations of the diet which would make sense, and obviously one would want to see that pursued. One would want to see efforts made, where there are obvious occupational factors, to remove them, although the general consensus now is that only a small part of cancer is connected with that. So, that's on the prevention side. On the cure side, the great problem and the obvious problem is that the chemicals—the drugs we now have to treat cancer—are non-specific, on the whole. They sometimes, for reasons that nobody is quite sure of, will be more antagonistic to the cancer cell than the normal cell, and that's why there are some remarkable successes in chemotherapy. But the general problem is the one of specificity, and that's what we're now trying to get at with the various approaches, one of which, of course, is the use of antibodies. The old idea of a magic bullet—which goes back to Paul Ehrlich before the First World War—is now becoming at least a foreseeable possibility. That is one area where a lot of effort is going and will continue to go, and where one would be surprised if there weren't some answers to come.

'Your reply was very applied. You wouldn't have asked the question "What's the *cause* of cancer?" It's almost as if that doesn't matter terribly.'

That's interesting, that you challenge me on that basis. I feel that the basic work goes without saying and continues, and that the need to find out the underlying causes is there and goes hand in hand with the applied work. I think we know a lot more about the origins and causes of cancer than is sometimes realized and said. A basic problem is really to find out what are the genetic functions that go wrong in the cancer cell and there are some very exciting directions to follow. So I think there'll be a lot of very important fundamental work there, but it will go hand in hand with the applied work.

'To what extent can you afford, since you are dealing with charitable money, to explore roads that just lead nowhere?'

Basic research is research which has no obvious goal in mind, which you do in order to understand further. And you don't ever know what part of it is going to be productive. I think it was the Nobel prizewinner John Vane, who used to be director of research at the Wellcome Foundation, who said to me, and maybe others have said it, that they know that 98 per cent of their research effort is not going to produce anything, but the trouble is they don't know which 98 per cent, you see! And, in the basic area that's even more the case. Obviously, you have general directions, and there are certain areas which you feel it makes more sense to support than others. Let me give one example. It's my belief that because of advances in tissue culture techniques and the use of the genetic engineering techniques, it's now much more possible to do work with mammalian systems than it was before. So other things being equal, I would feel there's some justification for supporting this exciting work using mammalian systems, as against using other model systems with fungi or lower animals. But we do have very good work going with *Drosophila*—that's the fruit-fly—genetics, and developmental studies. And you never know when they might not contribute to the mammalian side. There was the discovery of the so-called heat-shock proteins that are made in *Drosophila*. Well, there are heat-shock proteins in mammals, and they may be very interesting, and they will have been uncovered by the work in *Drosophila*. That's perhaps a rather narrow example of extension, but we need both. Also the ideas and even the techniques that are developed from the basic areas have their major impact on the applied areas. You do need both.

'You are interested in the relationship between science and society. How did you get involved in that?'

My original involvement came because at one time I was more of a population geneticist than I am now, and a population geneticist can have amongst his interests the genetics of behaviour, IQ, and things like that. Probably the most direct entry was through a concern about such issues. I wrote an article with an Italian colleague who is now in America, Cavalli-Sforza, on IQ and race, which tried to explain, as we saw it, the genetic basis for the arguments, and this made me a little more visible than I otherwise would have been in this area. This got me involved when I came back to this country in 1970, in chairing a group organized by the British Association for the Advancement of Science, that discussed these sorts of issues.

'What sort of issues?'

At that time—interestingly, because they're coming up again now— there was a lot of discussion about *in vitro* fertilization and cloning and things like that—genetic engineering. We also took in artificial insemination by donor, and transplantation. There had been a lot of talk about these issues, particularly their impact, and the thought was that the British Association could do something in this area by bringing together various people, and that I, as a geneticist who'd had some of these sorts of interests, might be a suitable person to bring together such a group.

'But when you say "bringing together", do you see scientists having a special responsibility in these areas?'

I think the responsibility of scientists lies in their knowledge of the science involved; when it comes to society's view, they have no more or no less a reason to put forward their view as members of society. But if they don't explain the scientific issues then nobody will. This BA group did not consist, by any means, only of scientists; for instance, it included a professor of theology, and a lawyer specializing in medical ethics. It was a mixture of scientists and others, including politicians.

'So it existed more to analyse and explain the issues rather than come to judgements about them?'

Yes, I think that's true. It was felt that the BA was a good group to bring the scientists together with others, raise the questions, and give a view as to what directions one might go with them. But not to prescribe.

'But do you agree that scientists have no more responsibility than any other citizen as far as the ethical decisions are concerned?'

Yes, I think that's true. As I said, I think their responsibility is through their knowledge. And if they don't, through their knowledge, bring these problems forward and explain what's going on, of course, then, I think in a way they're being deficient. Not every scientist has to do it. But there must be some who are prepared to explain and discuss. Because I think if you don't have that, you have a much worse situation—you have an ignorant public in the real sense of not knowing what's going on, making decisions inappropriately, often to the detriment of the science that the scientists themselves would like to do.

'Do you ever see anything in the argument that there is certain research one shouldn't do because it might lead to knowledge that is too dangerous to have?'

That's an argument that's often discussed, and the way you've put it, my answer is 'No'. In other words, I don't think that there is science that shouldn't be done because of the *knowledge* that is acquired. I think that what you do with the knowledge is for society to determine. I suppose fire has led to warfare, but it also does a lot of good things for us. What I do think, though, is that one can't say that there is never any research that shouldn't be done. There are obvious ethical constraints, for instance, on human experimentation, and there are controls in respect of safety and so on. So there are constraints of that sort which it would be ridiculous not to appreciate and abide by. In that sense there must be controls, but I don't think they should be in terms of the nature of the knowledge acquired.

'Do you think that there is something about this that has led to what is clearly an anti-science feeling?'

Perhaps I'm too much of an optimist, but I, personally, don't quite go along with this view that there's an anti-science movement. I think it only takes a few articles in the newspapers by a small group of people with a particular aim to give that impression. And I wonder whether, if you really ask the public in the proper sense of some sort of survey, they would agree with that view. I think there's still a lot of interest in medical research and what it can contribute. After all if there wasn't that sort of interest, how would we as a cancer research charity receive all the money that we do? The public has a confidence that there are things that can be done, and a trust in it. And when you see surveys—I know there have been some done on the other side of the Atlantic—of people's views of different professions, the scientists and medical people are rated quite highly. While there are natural concerns about what can be done— and they start with nuclear warfare on the one hand, and they go to tinkering with people's origins on the other—I'm not sure they really reflect a general concern and negative attitude about science. I hope not.

'I hope you're right too; but I think there is a feeling—and very often the media still present things this way—of the scientist very much as Doctor Frankenstein. The image is very hard to get rid of.'

I think it's unfortunate. I think there *is* a tendency to present the scientist as a bit remote, and part of it, of course, is that there's a difficulty in explaining science to the public.

'What is the nature of that difficulty?'

Oh, it's simply that there are words and concepts that most people haven't come across. And, if they don't understand, they may be somewhat afraid, and so they feel science is more remote. I know sometimes when I've done things about science on the media, you go to your relatives afterwards, and you ask what it was like, they say 'Terribly interesting . . . I really didn't understand any of it', even when you've tried to be very simple! But I believe you can explain anything given time and patience—and I think the news media often do an excellent job on that—but it's something which needs time and the care, and I think it's very important that it's done.

'Is it a tendency among scientists, *not* to wish to explain their work—as if it's not really *done* to appear on the media? If you're a young scientist, it's even been suggested it might injure your career if you're over-exposed. Do you think that's true?'

I think there's an element of truth in that. I think there are some scientists who take that attitude, and I sometimes find it, I suppose, amongst some of my colleagues. I think the important thing is that there are some scientists prepared to present science to the public who are respected because they're *scientists,* not because they're presenters, so that they can carry some authority in what they do. I think where you get into difficulty is when there are people who like to do this sort of thing who are not necessarily respected primarily as scientists, and they may tarnish the reputation of the others a little bit.

'You didn't think being a scientist necessarily lent a great deal to being an administrator, but I have seen you quoted as saying that when it comes to public policy decisions, scientists really have something different to offer.'

Yes, I mean that in this sense: scientists have a particular sort of training that I think is very important, and an approach to thinking about problems in a rational way which, I feel, can be applied in virtually all walks of life. I feel that there's a tendency for scientists almost studiously to be avoided, say, when it comes to government or administrative and executive offices in this country. There are traditions that somehow lead to the fact that a very small proportion of people who have had some sort of scientific training get into the upper civil service, or are members of parliament, for example; much smaller than their representation or importance in the community. I don't think they should go into these jobs because they're experts in a particular field of science, but they should go because from a scientific background they have something to contribute which gives a different point of view than say if you have come from an arts background or a legal background. I think that should be represented in government much more than it is—government in the broadest sense of the word.

'But, could you just explain why, apart from a special knowledge of technology, say of nuclear reactors, scientists have anything more to contribute in terms of their training than someone who was trained in history or in any other scholarly discipline?'

Well, my first answer would be, I don't think they necessarily have anything more, but they certainly don't have less, and at the moment I bet you'd find fewer scientists in the upper civil service than historians. So my first thing would be to redress the balance until it's at least equal. I think my second point would be that I do think they have something that is different because of the logical, systematic sort of training that one gets in science. I'm not saying that it's superior, but it is something different. It's a way of thinking; even, if you will, of dealing with symbols if you've been trained in the physical sciences; of not being afraid of some of the terminology and therefore being more willing to go from one area to another. I think that those people with that training in science who are then willing to go out and do other things do have something quite positive to offer.

CUNNING
MECHANISMS

RICHARD GREGORY *was born in 1923 and is Director of the*
Brain and Perception Laboratory at the
University of Bristol.

CHAPTER 14

CUNNING MECHANISMS

—◦◦◦◦—

RICHARD GREGORY

Neuropsychologist

THE FIRST few moments of the Universe are discussed with greater certainty than the human ability to think about such problems. Psychology is hardly recognized as a respectable science, except when it deals in the 'hard' scientific concepts of neurophysiology or contemplates artificial rather than human intelligence. Yet the nature of mind, like the origin of the earth and the stars, has perplexed human beings from the very beginning. Richard Gregory, Professor of Neuropsychology at the University of Bristol, is one of those trying to make some progress in thinking about thinking. How does he go about it?

In Martin Escher's famous lithograph *Ascending and Descending* monks trudge endlessly up stairways and yet always find themselves back where they started. For Gregory, such optical illusions are not just a source of great intellectual pleasure, but of serious scientific interest. They represent errors in perception, and so provide a glimpse of the way in which the brain makes sense of the signals it receives through the eye. He is also concerned with the insights the use of computers can provide into how the brain does—or does not—think. One of his heroes is Alan Turing, who, though he died in 1954, laid the foundations of artificial intelligence. Turing asked whether the mind of a human being could adequately be described by analogy with a computer that simply manipulated long strings of symbols. He suggested that one could judge the success of a computer simulation of the human mind by whether an interrogator who could test only mental attributes, such as the ability to play chess, could determine whether he was dealing with a machine or a person. The problem at the heart of artificial intelligence is whether the most successful problem-solving machines—such as the digital computers which, following logical or arithmetical rules, can play chess as well as, or better than most people—in any way reproduce the mental processes of their human opponents.

An experience that deeply affected Richard Gregory was the case of a man who had been blind from infancy but gained his sight in middle life. Gregory found that the man was immediately able to see things he was familiar with through touch, but was for a long time effectively blind to other things. Not only did this suggest certain things about the way in which we perceive the world, but it was also one of the events which eventually led Gregory to set up the Exploratory in Bristol. This is a science museum which encourages people to discover the laws of nature through 'hands on' demonstrations and apparatus of all kinds.

Gregory is a large man, and his ebullience makes him seem even larger. He has a penchant for puns and takes enormous pleasure in machinery. In the sitting room where we sat to talk there was a handsome brass telescope at the window, and on a side table a toy robot stood next to one of the earliest mechanical calculators. As a setting for the contemplation of grand problems it had a delightful—if slightly deceptive—lack of gravitas. Inventor, psychologist, neurophysiologist, and popular entrepreneur . . . I wondered how Gregory would wish to define his diverse and highly personal brand of reflexive science.

———

I think I see myself as trying to be an experimental philosopher, which sounds a bit arrogant, but I'm interested in the relationship of the mind to the physical world. I'm interested in perception because I think of us as active processors of information, making sense of the world somewhat as scientists do with data. I'm a *psychologist* in that I'm interested in the procedures that the brain is handling; and I think *physiology* is absolutely great because it tells you about the mechanisms and how they arose in evolution.

'But why a philosopher? I don't see where the philosophical side comes in.'

I think one needs to be a bit of a philosopher in this sort of subject because one is trying to relate very different kinds of experiments. For example, one tries to relate the physiological data, as recorded by electrodes attached to the brain, to what the signals actually mean, in terms of how the brain makes meaning from the little pulses of electricity running down nerves. This is a tremendous problem. I think it's a *philosophical* problem how signals are read as messages.

'I'm surprised. I think most scientists would really deny that they were philosophers. It does sound like a difficult problem, but what is it that *philosophy* gives you that science itself doesn't?'

Well, I think that there are certain questions about the status of data. For example, what is a signal? Let's take a computer. You say it's got both software and hardware, but what is the exact relationship between the two? This I think is a conceptual problem because when the computer goes wrong one can say that the program—the software—is not appropriate to the problem, or one might say that the electronics have gone wrong. These are very very different accounts, very different theories, and I think this relationship between a mechanism and the function of a mechanism—what it's doing, what it's handling, the procedures it's carrying out—is a difficult problem, because it's not quite within science. I think you've got to think about it very, very generally, and I call that philosophy.

'What do you mean "not within science"?'

I don't mean that it is something which is outside science like a balloon hovering around outside what can be known by science. I see philosophy as sort of proto-science, getting the questions asked and initial demarcations set out. Then one goes to the nitty gritty, does experiments, develops techniques, then refines it, and very often changes the philosophy completely. But I think you need it as a starting point.

'Were you trained as a philosopher?'

Yes, I did two years of philosophy at Cambridge. In fact I was supervised by Bertrand Russell for a time. When he was 76 he came up to Cambridge one day a week and supervised six of us. Then I did experimental psychology, and I dabbled in physiology.

'Had you always been concerned with the mind? I mean at what age did you decide that you really wanted to work on the mind?'

My father was an astronomer and I attribute it to that, to looking through telescopes when I was quite young. You see a little flickering image, and then you look in the book and you learn what's actually *there,* which is incredibly different from what you see. I think really it's that problem of how you read the world from flickering images, from indirect data, which has always rather fascinated me.

'When you say flickering you mean . . . ?'

Well it's literally flickering, of course. If you look through a telescope the atmosphere actually disturbs the image. It is really flickering like a flame most of the time, particularly in English weather. My father's observatory was in London and the turbulence was pretty strong. So what you see really is moving around although the object is, of course, static, or still, and you're sampling it. And this, indeed, is a problem which concerns me. I actually designed a telescope camera for minimizing the effects of this disturbance by the atmosphere. This started me thinking about what it is to see an object when you see it from all different points of view. Often bits are hidden, so how do you construct the object in your brain or in your mind? That itself, of course, is a philosophical distinction. Then, what are the physiological mechanisms analogous to how you do it electronically or with telescopic optics?

'Are you particularly interested in technology?'

Oh very much so, yes. I think technology provides both the questions and the answers to all sorts of deep questions because we think in terms of models. I do actually think that one needs, in the sort of field I work in, to try to encompass the basic principles of technology and something of its history, because these are our thinking tools. In the seventeenth century, the nerves were thought of as tubes carrying a sort of vital spirit. Well, the idea of tubes and fluids running down them was the technology of the seventeenth century, when garden fountains and pumps were first invented. Then, of course, they've been thought of as telephone exchanges, and then computers. All these are drawn from technology and provide models for thinking.

'Presumably the paradigm that you would use now for thinking is computers?'

Yes. I'm in a bit of a puzzle about this one, actually. One thinks now of a computer as a digital computer which goes through perhaps thousands of steps in order to solve a problem. If it's, say, differentiating, it goes through the steps of an analytical calculus and it calculates the answer. Now I don't know, but I somehow feel it's a bit unlikely that the brain actually works quite like that. I don't believe that, so to speak, it knows advanced mathematics in order to pick up a coffee cup; I somehow don't believe it. I think it's done in a cruder way, and I think we just haven't got the technology yet to give us the right answer.

'You mean to give us the right model for how to think about the brain?'
Indeed yes.

'But I want to understand better the problem that you're thinking about: which problem do you really want to solve?'
I think there are very general deep questions one probably never will get answered. And then, of course, there are certain particular, rather technical questions, that one plays about with all day long. But I think the sort of general question is the hoary old one about the relationship of the mind to the brain and, indeed, what the word 'mind' really means. We know what 'the brain' means—we've got a physical lump in the head which we can touch; the mind looks sort of ghostly.

'But what do you mean by "mind"? Do you mean what we can feel . . . love and hate?'
Supposing one says one is doing mental arithmetic . . . in what sense is the arithmetic being done by mental processes rather than brain processes? One could put it perhaps that way. But if one feels pain or love, it's very difficult to imagine a machine that can feel such things, even though you can imagine a machine that will do arithmetic. This is where the problem takes off I think.

'Now do you want to understand this clearly very deep problem because it will solve all sorts of other problems, or because you just think it's so important?'
I think probably the second. I just think it's extraordinary that human beings have been questioning this right through recorded literature. Homer, for example, thought very much about it. In mediaeval philosophy there are all sorts of religious answers; at the moment, it's a question that ought to be answerable by science, by physiology and psychology. But even so I feel that the problem remains: how do we state the question before we can do the right experiments? And therefore it's still a philosophical problem, I think.

'But when you approach this problem do you set aside a part of the day and say I am going to think about thinking?'

I wrote a book called *Mind in Science* and I did a lot of thinking about the problem while I was writing that over ten years. And then I write editorials for the journal which I edit, and I use that as a kind of stimulus to think about problems. It might be a jokey, trivial sort of problem, or it might be a technical problem, or yes, it can be this sort of problem, which I actually very much like to think about. I find it very great fun.

'So do you find that your way of thinking is actually by writing?'

Yes. I use writing as a kind of scaffolding for thinking, very much.

'Did you enjoy writing you ten-year book?'

Yes. It was a bit like living with a tempestuous woman in a way, who sometimes quarrelled with one and certainly affected one's life very deeply. It's very much like a human relationship really, and I kept going back to it and having conversations with it and learning from it and teaching it, and so on—just like a human relationship.

'Did it take on an identity of itself'

Yes. I mean, I kept re-writing it; I never had a very exact plan. And of course it made me read. I read practically the whole of Aristotle. I wanted to find out how people had thought in the past. What I found, which surprised me, was the importance of technology. The Greeks had an awful lot more technology than most classicists give them credit for—quite frankly, I think, because almost all classicists have no interest in technology themselves. They don't know one end of a bicycle from another. So they sort of filtered out what seemed to me incredibly interesting—that the Greeks actually made wheeled computers, not for doing arithmetic but for solving the problem of the movements of the planets. One of these actually exists—it was found at the bottom of the sea and is a lump of bronze which was made in 80 BC and is itself a child of many generations of previous wheeled computers. Now it's astonishing that the Greeks had this. There are plenty of references to these things as well as the actual lump of bronze, and clearly they were interested in technology.

'You obviously get a great deal of pleasure from technology. If I look around this room you have wonderful old microscopes, a telescope at the window, and a little toy robot. What is the pleasure you get?'

Well, I think it's at many levels. I think any example of human skill is obviously interesting. The development of ideas as embodied in machines is just as interesting as the development of ideas embodied in books, and in many ways books are surrogates of what people are actually doing. I think one can only understand books when you know something of what the author has experienced or what he's made or what he's used in his lifetime, because I think our experience comes primarily from handling objects, not words. So I think in a way you get to the springs of experience by looking at activity—not symbolic activity either, but actual activity—solving problems, making things, doing things.

'Are you good with your hands? Are you a designer of machines? Can you invent things?'

Yes, I'm a bit of a gadgeteer, rather than an engineer. I hold about 30 patents of various kinds. I do invent things, yes.'

Like what? What have you invented?'

I invented a microscope working in three dimensions, which I call the solid image microscope. That was a rather fun instrument.

'Does anybody use it? Has it been made?'

Not very successfully. It has been used, but there are problems because when you put mechanically vibrating parts on to a microscope the wrong bits start to vibrate. It's really been superseded now; there are better ways of doing it. But, interestingly, exactly the same idea is now being used in computer graphics to get three dimensions with a computer picture.

'But what do you see as the relationship between inventing things and doing science? Do you see them as very similar activities?'

I think I do, because I believe that thinking and problem solving are active, and I think that intelligence is an activity. It's finding novel solutions. 'Appropriate novelty' is how I would define intelligence. So I think inventing—not necessarily some dramatic invention like the telephone, but inventing on a smaller scale—is what organisms do whenever they're intelligent. Therefore the process of inventing is to me very, very interesting.

'You often refer to computers. It almost seems that computer technology is part of your language. When you wrote your book did you use a word processor?'

As a matter of fact I didn't have my word processor until I'd finished the book, and I would probably have written it about three years sooner if I had. In fact I bought the word processor just after finishing my book.

'Do you like using your word processor?'

I love it, yes. I think one can have a relationship with a machine that you interact with, such as a word processor, or a car, almost in human terms. I get quite fond of the machinery around me, particularly when it's of an interactive sort.

'A lot of people really are addicted to word processors. I was at an Oxford college recently, and the only conversation at lunch amongst the arts people was about various technical aspects of their word processors. Where do you think this passion comes from?'

I suppose it's something like the master/slave relationship. I mean, they're infinitely obedient, they don't answer back—like a good secretary really. I think there are two kinds of secretary that are very good: the very, very reliable, totally submissive ones who find what you want in the filing cabinet, but actually get a bit dull, and the other sort who steer one, and say for God's sake get on and write such and such a letter and so on. Now the second sort are actually much more fun to work with, but the word processor is so obedient that it pushes that first, optimal kind of secretary to the extreme limit, and is a kind of perfection.

'I would like to ask you about psychology in general. Would you say that psychology really has made the sort of strides that we've seen, for example, in the biological sciences? Has there been anything analogous to the discovery of DNA, or is it a subject that's still waiting for some such discovery?'

That's a very interesting question. My own view is that we're still waiting. In my view technology is necessary for pure science to develop its models and its tools, and I think we're actually waiting for, as it were, cognitive computers. That is, computers that not merely know grammar and logic, but can understand the meaning of sentences, and the meaning of signals coming, let's say, from a television camera eye feeding into it. If it can actually understand 'it is a table', 'it is a bottle of wine', and then *know* what tables and bottles of wine are for and can do, then I think we're going to start to get a technology which is going to be adequate for thinking about psychology. We're really waiting for that. At present we're still messing about with inadequate models, and I think too much is claimed of computer models at the present time, because computers really are pretty dumb and they don't understand the significance of the symbols they handle.

'A lot of your work has been in the perception of visual illusions. Why did you choose that particularly?'

I like illusions for the following reason. I think that science has always done very well when it's got nice phenomena to look at, and the more mysterious and perhaps trivial looking the phenomenon to start with, the better. Let me give you an example—the lodestone. The Greeks were fascinated by lodestones and they first of all gave them a psychological explanation. They said they were attracted to each other as men and women are attracted and, in fact, the dark ones were called male, and the lighter stones were called female. And then gradually this sort of psychological explanation gave way to physical models, but with still a great mystery about how two objects can attract each other without a piece of string between them. They couldn't see what the force was. I think this extended people's thinking to kinds of explanation that would describe repeatable phenomena, even though the source or cause or mechanism of the phenomena was not visible. Now the interesting thing about illusions, to my mind, is that they are phenomena—they're repeatable. But they're not phenomena of physics, because they're distortions from what we know is actually present. Therefore I regard them as phenomena intimately connected with mind or brain—I won't make the distinction at the moment—but they're a kind of unnatural phenomena rather than natural phenomena. And to find out just why it is that, in a certain simple shape, a line looks too long or it looks curved, is something you can do proper science on. It's something you can look at in different situations; you can make measurements; you can have controlled conditions. And you can set up hypotheses—guesses—as to what has gone wrong in the mind or in the brain of normal people who are thoroughly abnormal in that particular situation. Also, it's a way into the relationship between physiology and psychology. Is it that the nervous system can't somehow represent that shape in terms of its normal signalling, or is it that the signals are not read appropriately? You've really got two kinds of illusion: you've either upset the mechanics of the system, or you've upset how the signals are read. To put it another way—going back to the computers—is it a hardware error or is it a software programming error? The interesting ones are the second kind.

'It's really rather fun dealing with illusions. I mean they are a little mysterious. They are a bit like a trick. Was there any quality of that which appealed to you?'

I don't think so, not directly. What I had to do, actually, was to see that although they look trivial, they're not trivial. When you look at an illusion in a children's book—distorted lines or Necker cubes which reverse, something like that—at first you write it off as trivial. The effort, intellectually, is to see that underneath that, it needs an explanation. I got involved in space research when I was at Cambridge, and I ran a group for the American Air Force. They were really interested in the problem. For them it was a serious matter whether astronauts would suffer illusions when they were docking space ships or when they were landing on the Moon, because the conditions are pretty atypical. So I built a space simulator—on American money I may say! We enjoyed that, we had tremendous fun. The trouble is we spent so much time building the simulator that we didn't do very many experiments on it, but it was tremendously amusing.

'Did it help with the landing on the Moon?'

Indirectly I think it probably did.

'Are you very visual? Do you care about painting? Has that anything to do with the choice of illusions?'

No, it was intellectual. I'm very, very bad, actually, at drawing and painting. I quite like painting, but I prefer music. I'm not really particularly visual in that sense.

'Well, the reason I asked that was that I know that you're terribly fond of puns, and it did strike me that puns are a sort of verbal illusion.'

Yes, I think that's right. I think it's the underlying characteristics of the suddenness, the surprise of illusions that appeals to me, rather than their aesthetic character. I think they're not particularly beautiful, but they're terribly intriguing to my mind, rather like puns, or cunning mechanisms.

'So there's a link with the cunning mechanisms of the objects we see around the room.'

Yes, but mechanisms of thinking as well.

'Is that perhaps really why you're interested in mind and brain—because they're so surprising, rather than mysterious?'

Oh absolutely. I don't like mystery for its own sake. I like puzzles. I don't like games so much because I think, frankly, one's rather wasting one's time. I used to play chess a bit, but I now think I'd rather try to think out a puzzle in my experiments or in my philosophy than play a game of chess. But it's got that fun about it. It's playing a game either against one's colleagues or somebody who's written a book you think might be a bit silly or, perhaps, wiser than oneself; but particularly it's a game against nature, against the way things are. And you try and win—there's a certain competitive streak here, I think—against reality itself.

'So while the computer is the perfect slave, it's more fun to play against nature which is much more devious and cunning.'

That's absolutely right, and I must admit that computer simulations, rather as pictures, get a bit boring. One wants to contact as close as one can the world itself and find out its surprises.

'Is the pleasure that you get, then, from the surprises rather than the solutions? I have the feeling that once you've solved a problem you really find it a little dull.'

Yes, that's quite true. I often find it an effort to write the thing up and publish it once I think I know the answer. I prefer writing when I nine-tenths know the answer because then I find the activity of writing part of the problem solving, and then I find it's very stimulating and fun to do.

'So you're not terribly interested in adding to the general body of
knowledge? Isn't it rather self-indulgent that you have a lovely position
where you're playing this fun puzzle game against nature for your own
gratification?'

Yes, you're absolutely right; it is a very valid criticism.

'But it wasn't meant as a criticism.'

No, it's a valid criticism. I think it is self-indulgent actually.

'Why I said it wasn't a criticism was because I feel quite a lot of my own
work sometimes is of that class. I too enjoy the puzzles. But unlike you, I
think I get great pleasure from telling other people if one has got a
solution. Isn't there the joy of telling people that you've won, as it were,
against nature?'

Yes, well, perhaps you are lucky. You see, most of my colleagues don't
believe in my solutions . . . you may be more fortunate.

'Do you enter into battle with them?'

Yes. I've had a bit of a running battle about illusions, funnily enough, for
over twenty years. It really is very amusing. You see, I think that
physiologists want to own them, in a way. Now, unless the explanation
lies in the sort of models, or experiments, or way of thinking that you
do, you don't own the phenomena. So there's a bit of a rivalry here
between psychologists, who are concerned with the processes of seeing,
and the physiologists who are concerned with the mechanisms by which
the processes are carried out. So if it's a mechanism that has gone funny
or wrong, as with overloading, or some such thing, or after-images,
which are clearly purely physiological in that sense, then they belong to
physiologists. But I happen to think that some of these phenomena have
to be understood in terms of the procedures of perception. If one
understands the phenomena, then one owns them. So I want to capture
them into my territory, and there really are territorial battles between
physiologists and psychologists; we've had this for many years.

'How are these battle fought?'

Not altogether fairly as a matter of fact. They're very much fought by
forgetting half of the counter-evidence, of course. This I think is true in

the whole of science. There's actually quite a lot of slightly dirty play that goes on in a way because, in a sense, this is right too. I mean, if you're pushing forward with an idea and you think of all the objections or all the counter-experiments that look as if they go against it, you don't actually get anywhere. In a certain sense you have to steam ahead and say 'Oh well, I'll solve that one when I get to it, so let's forget it for the moment,' in order to make any sort of progress. Certainly in Cambridge people are very, very tough, and if you give a talk or you present an idea which is not fully worked out—and of course it can't be fully worked out while it's in the embryo stage—it can get killed as an embryo before it ever happens. So I think some tolerance both to oneself and to one's opponents when they can't quite answer a question, or overcome a difficulty, is very necessary, actually, to make progress.

'When that happens to your work what do you do about it?'

Well it depends on one's mood. It can actually be quite upsetting in a way—it really can be—when you've spent a lot of time and you do an experiment which you think is really neat and nobody can be bothered to look at it because it doesn't fit in with how they think. It can actually be upsetting.

'You're deeply involved with popularizing science at the moment?'

Yes.

'Why?'

Well, I've always in a way liked to be, not just at the moment. I wrote a book called *Eye and Brain* nearly twenty years ago, now, which is in umpteen languages, and it's read widely. That is, I suppose, a popular book isn't it? It tries to present visual perception to artists and students of psychology, physiology or indeed philosophy, and it's read by architects and so on as well. I just think it's like sharing a toy or sharing something that one likes . . . or it's like giving a party. I think it's a very basic thing, sort of 'come up, and let's have fun with this thing'. And of course it reflects back, and you have far more fun yourself because you're kicking it around with other people. This, I think, is what science should be like—not working in cupboards and getting amazingly aggressive about other people who think a bit differently, but treating it as a tremendously exciting game; getting involved in it in the way that you do with a game, but maintaining friendly terms with the opposition. I think when you popularize it, it really is like inviting people into your party!

INDEX